Charles Darwin's Barnacle and David Bowie's Spider

CHARLES DARWIN'S BARNACLE AND DAVID BOWIE'S SPIDER

HOW SCIENTIFIC NAMES CELEBRATE ADVENTURERS, HEROES, AND EVEN A FEW SCOUNDRELS

• • •

STEPHEN B. HEARD

WITH ILLUSTRATIONS BY EMILY S. DAMSTRA

Yale
UNIVERSITY PRESS
New Haven and London

Published with assistance from the Louis Stern Memorial Fund.

Yale University Press books may be purchased in quantity for educational, business, or promotional use. For information, please e-mail sales.press@ yale.edu (U.S. office) or sales@yaleup.co.uk (U.K. office).

Set in Gotham and Adobe Garamond type by IDS Infotech Ltd., Chandigarh, India.
Printed in the United States of America.

ISBN 978-0-300-23828-0 (hardcover : alk. paper)
Library of Congress Control Number: 2019947766
A catalogue record for this book is available from the British Library.

This paper meets the requirements of ANSI/NISO Z39.48-1992 (Permanence of Paper).

10 9 8 7 6 5 4 3 2 1

Contents

CONTENTS

Preface

Part of what makes us human is our curiosity about the world around us. That curiosity drives scientists to discover, describe, and name the millions of other species with which we share our planet. Every now and then, the naming of a newly discovered species catches the public eye. Sometimes, it's because that new species is named after a person—be they living or dead, real or imaginary, admired or loathed. Among such eponymous namings are the barnacle named for Charles Darwin (*Regioscalpellum darwini*), the spider named for David Bowie (*Heteropoda davidbowie*), the fungus named for Sponge-Bob SquarePants (*Spongiforma squarepantsii*), and the beetle named for George W. Bush (*Agathidium bushi*). These namings, and many more like them, connect the scientists who name species, the species that bear the names, and the people to whom the names refer.

They also strike many people as a little peculiar. What a strange way, it seems, to pay tribute to someone—to build their name into a quasi-Latin name for a species, a name that will be used primarily by scientists writing technical and jargon-filled journal papers and monographs. You can almost sympathize with Jane Colden's aunt. Jane Colden was likely the New World's first female botanist. She was active in the middle of the eighteenth century; her father (the marvelously named Cadwallader Colden) was also a botanist and supported her interest in natural history. Jane's handwritten and hand-drawn manuscript illustrating the flora of New York circulated in London, and it was suggested that a plant should be named *Fibrurea coldenella* in her honor. Her aunt, however, was shocked, and objected, "What! Name a weed after a Christian woman!"[1]

It was Carl Linnaeus, the brilliant eighteenth-century Swedish botanist, who made it possible for a spider to bear David Bowie's name (or for a plant to bear Jane Colden's)—and thus, for a name to tell a story. Before Linnaeus, the scientific naming of a plant or animal species was simply an exercise in description. A name was a Latin phrase (sometimes quite a long one) that described the species and separated it from similar ones; but it did no more than that. Linnaeus's "binomial system" was different in several important ways. Most celebrated is that it's simple, and allows easy organization of our knowledge about the Earth's biodiversity. Each species has a single-word name, coupled with a single-word "genus" name for its group of immediate relatives—*Acer rubrum,* for example, with *rubrum* designating one of 130 or so living species of maples in the genus *Acer.* But a less widely appreciated novelty in Linnaeus's system was its separation of naming from description. Linnaean names—and all scientific, or "Latin," names since Linnaeus—are indexing devices. They may be descriptive (as in *Acer rubrum,* red maple), but they don't have to be (as in *Acer davidii,* Père David's maple).

Linnaeus's invention of non-descriptive names might seem trivial, but it made something possible that had never been possible before: in naming species, scientists could express themselves. In choosing to honor someone with an eponymous Latin name, a scientist can tell a story about the person being honored; but at the same time, that scientist tells a story about him- or herself. With Linnaeus's invention, names—and eponymous names in particular—became a window on scientists' personalities.

What shows through that window? That scientists aren't the cool, dull, and unemotional creatures that many might expect. They use Latin naming creatively, and in doing so reveal all of humanity's virtues, weaknesses, and foibles. In naming organisms, some scientists proclaim their admiration for naturalists, explorers, and others among their heroes. Some acknowledge their gratitude to their mentors or patrons; some express their love for their husbands or daughters or parents. Some stake their claims as fans of Harry Potter or of punk music. Some make statements about justice and human rights. Some signal their disdain for demagogues and dictators; others, sadly, reveal their approval of them. The eponymous names of organisms can reveal the shame of bias and prejudice, but also the pride we can feel in attempts to rise above those human failings. In coining eponymous names, scientists show themselves as sometimes sober, sometimes playful, and sometimes eccentric; sometimes gracious and sometimes spiteful; and every bit as passionate about history, arts, and culture as they are about the pattern of scales on the belly of a snake.

Through the window of eponymous naming, we see the best and the worst of humanity. We see science as a fully human activity, full of personality and history and shaped by intriguing connections between the species that's named, the person it's named for, and the scientist doing the naming. As Mistress Mouffet puts it, in A. S. Byatt's novella *Morpho Eugenia:* "Names, you know, are a way of weaving the world together." The stories woven by eponymous names can be surprising, fascinating, and poignant; occasionally, they can be infuriating. The pages that follow tell some of those stories. Enjoy your look through the window.

Charles Darwin's Barnacle and David Bowie's Spider

INTRODUCTION

A Lemur and Its Name

You and I are primates. Our lineage, among mammals, stretches back 75 million years. In some ways, it hasn't been a terribly successful one. Only 504 living primate species are known to science, compared with (for example) 1,240 species of bats, 26,000 of orchids, and 60,000 of weevils. At the same time, though, it's a lineage that—through the actions of our own species—has transformed the planet as no other ever has. We can be both ashamed and proud of this. It's true that no species has ever polluted so many lakes, razed so many forests, or driven so many other species near, or to, extinction; but it's also true that no species before us has written symphonies or built libraries or art galleries. Even more: no species has thirsted so much to understand its world, and no species has ever made so much progress toward that goal. Like every species, we have our local preoccupations with territory, food, and mates; but we've also looked up from them to study and name rocks, plants, animals, landforms, even planets and stars.

As human beings, we tend to be intensely interested in our own kin. That's true in genealogical and geographic senses: we cherish our families and our local communities. But it's also true in an evolutionary sense. The nineteenth-century discovery of our close relationship to the other great apes sparked a public controversy that continues in some circles today. The primate house is a showpiece attraction at any zoo, and we devour articles and documentary films about chimpanzees, gorillas, and our other close relatives. And when I take undergraduate students on field courses to the tropics, nothing excites them more than a chance to glimpse a monkey swinging high in the forest canopy.

Surprisingly, perhaps, our knowledge of our primate kin is far from complete. We know quite a lot about chimpanzees, bonobos, gorillas, and orangutans, which are the largest primates and our very closest relatives. We know much less about the rest of the group. A few are both well studied and staples of popular culture—think of the Japanese macaques that make such captivating images when they bathe in hot springs in winter. Most others are understood superficially at best. Deep in the forests of Madagascar, for example, live primate species newly recognized by science and virtually unknown.

Madame Berthe's mouse lemur,
Microcebus berthae

Among these most mysterious of our relatives are the mouse lemurs. There are 24 species of mouse lemurs in Madagascar; all but two were unknown to science just 25 years ago. One of these newly recognized species is the smallest of all living primates: Madame Berthe's mouse lemur. A full-grown adult would fit comfortably in the palm of your hand, and weighs just 30 grams—about as much as five U.S. quarters, or a slice of bread.

Madame Berthe's mouse lemurs live only in a small area around the Kirindy Forest on Madagascar's western coast. Kirindy is a tropical deciduous forest—sparse and quiet through the long dry season when the trees drop their leaves and much of the fauna hunkers down to wait out the drought. When the rains return, the forest becomes a dense tangle of green and a hive of activity. If you visit Kirindy as the rainy season begins, around dusk you might feel a breeze hint at relief from the day's heat while the last traces of sunset pink fade in the western sky. Stay still for a while, and you may hear rustling in the branches, and perhaps a soft chittering call: mouse lemurs are emerging from their daytime treehole nests to forage in the forest understory for fruits, tree sap, and insect honeydew. A well-aimed flashlight may catch some pairs of curious eyes, reflecting a softly gleaming orange. The smallest pair of those eyes belongs to our smallest of kin.

We've known for a long time that mouse lemurs live in Kirindy Forest. Only in the mid-1990s, though, did scientists realize that Kirindy mouse lemurs were not one, but two species. The larger one, the gray mouse lemur, has been known to science since the eighteenth century; but the smaller, Madame Berthe's mouse lemur, was formally recognized and described by scientists only in 2001. In the process, it received a formal ("Latin," or "scientific") name: *Microcebus berthae.* That name, it turns out, honors a woman named Berthe Rakotosamimanana. Who was Berthe Rakotosamimanana? How did the tiniest of all our relatives come to bear her name? What did she do to deserve this tribute? For a tribute it is—an odd kind of tribute, it might seem to many, but a heartfelt tribute nonetheless.

Everyone knows that science can be dull and fusty, and that the Latin names we give to plants and animals are the dullest and fustiest of all. They're long, they're unmemorable and unpronounceable, and they're at best a necessary evil that biology students memorize as some kind of scientific hazing ritual. Everyone knows this. But everyone is wrong. Sure, some Latin names are difficult or obscure; but others are marvels. In the pages that follow, I'll share with you the stories that lie behind Latin names honoring people—explorers, naturalists, adventurers, even politicians and artists and pop singers. These stories are a window on the culture of science and the personality of scientists, and they reveal fascinating connections between the scientists who name species, the people they honor in the naming, and the organisms that bear the names. We'll return to the story of Berthe Rakotosamimanana and her mouse lemur in the Epilogue; but in the meantime, there are plenty more to tell.

1

The Need for Names

"What's the use of their having names," the Gnat said, "if they won't answer to them?" "No use to THEM," said Alice, "but it's useful to the people that name them, I suppose."

—*Lewis Carroll,* Through the Looking Glass

Our planet teems with life. Tropical rain forests and coral reefs are staples of documentary filmmaking, in part because they offer astonishing biodiversity: wherever you turn, a new and different species comes into view. A football-field-sized piece of Amazonian rain forest might have 200 different species of trees—and that's just the trees, which are far surpassed in diversity by herbaceous plants, by insects, by fungi, by mites, and by members of many other groups. Visit an Indonesian rain forest, or even just a different tract of Amazonian forest, and you'll find different species; leave the rain forest for a dry forest or a cloud forest or a savannah, and you'll find different species again. This pattern continues as you travel the globe: some habitats are richer in life than others, but all add to the tally of species on Earth. Even habitats that seem utterly hostile to life have their denizens: boiling hot springs, the deepest caves, the snowfields of the Himalayas, cracks in rocks a kilometer underground.

How many species share our planet? For biologists, it's both exciting and professionally embarrassing that we don't know. We don't even have a particularly good guess; or rather, we have a lot of guesses, but they don't narrow the range much. We do know that it's a lot. Somewhere around 1.5 million species have been formally described and named by science; most estimates of the total number of Earth's living species are somewhere between 3 million and 100 million. A recent estimate of a *trillion* species just of bacteria and other microbes raised a lot of eyebrows. Although that number's plausibility has been hotly debated, its publication makes it obvious that we can't put a clear limit on how high the count could go. But each of these estimates counts only *living* species. Over the four-billion-year history of life on Earth far more species once lived but are now extinct, and this fact makes the planet's biodiversity even more astonishing. If extinction has claimed 99 percent of all species that have ever lived (itself a guess, with the real fraction most likely higher), then all-time species counts are two orders of magnitude higher than our counts of species alive today. 300 million? 10 billion? *100 trillion?* Each is (or was) a distinct species with its own morphology, behavior, habitat preferences and requirements, and ecology. This is mindboggling, wonderful, and something of a problem.

Why do I say Earth's diversity is something of a problem? Because all of these species need names. They need names for psychological reasons, and they need names for practical ones.

Psychologically, naming gives us comfort with the existence of Earth's species and allows us to make progress thinking about them. Actually, this is true not just of living organisms, but all the other kinds of entities we name. The brilliant French mathematician Alexandre Grothendieck, for example, wrote that "one of my passions has been to give names to [mathematical concepts] as they reveal themselves to me, as a first step in understanding them."[1] Grothendieck was well known for the care he took in naming new concepts or mathematical objects, in ways that brought attention to them and shaped the way people thought about them. Something similar was involved in Georg Cantor's discovery that some infinite sets are larger than others (in particular, some are infinite but countable, while others are uncountable). Cantor gave these different sizes of infinities names (assigning them what we call aleph-numbers). By naming infinities, Cantor in an important sense made them accessible to mathematicians and to further mathematical thought (and in doing so he sparked a storm of mathematical controversy). What's true for abstract mathematical concepts is no less true of concrete things. It just feels natural to talk about something that's named, but difficult and vaguely discomforting to talk about something that's merely described.

Naming seems to be deeply seated in who we are as a species. In the Old Testament creation story, naming Earth's creatures was Adam's first job: "And out of the ground the Lord God formed every beast of the field, and every fowl of the air; and brought them unto Adam to see what he would call them: and whatsoever Adam called every living creature, that was the name thereof" (Genesis 2:19, King James Version). For good or bad, naming may even make us feel like we have some power over the things we name. From Egyptian and Norse mythology to *Rumpelstiltskin* to *A Wizard of Earthsea,* our stories touch on naming, and power in naming, over and over again.

Even in the absence of our psychological compulsion to name, though, the practical need for names would dictate that we do so. That's because we need to keep track of our planet's millions of species. We need to be able to talk about them, and to know which one we're talking about unambiguously among all of the millions of possibilities. After all, "the red berries are good to eat, but the blue ones are poisonous" breaks down pretty quickly in a world as diverse as ours. If legislation prohibits development of the habitat used by a species that's at risk of extinction, both developers and conservationists want to be sure that we all agree on which species that is, so we can identify those off-limits habitats with precision. If we've discovered a promising new cancer drug in a deep-sea sponge, we need to be able to tell each other which sponge it is, so that we can all be testing the same extract. If a child has been poisoned by a mushroom, we need to be able to tell doctors which one, and those doctors need to be able to look up symptoms and treatments for that precise mushroom.

Naming solves the keeping-track problem: each distinct entity to be kept track of gets a name, and that name works both as a label by which we refer to the entity and as an indexing mechanism by which we tie the entity to the knowledge we have about it. We use names to keep track of the children in our families, the minerals in Earth's crust, automobile models in showrooms, and listings on the stock exchange. We do this with living species, too: we name grizzly bears and polar bears and spectacled bears and panda bears; coho salmon and Chinook salmon and king salmon; tulips and geraniums and daffodils.

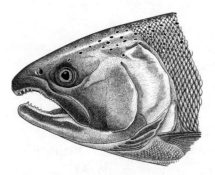

Coho salmon, *Oncorhynchus kisutch*

Grizzly bear, coho salmon, and *daffodil* are examples of common names: the names we use routinely in everyday conversation. Common names can be fascinating and evocative, with origins in description (the fork-tailed *swallow,* from an Old English word for a cleft stick), onomatopoeia (*rook*), folklore (*goatsucker*), and even eponymy (*Darwin's finches*). But for a variety of reasons, they fall far short of doing the job we need naming to do. For one thing, not infrequently, common names are inaccurate: Darwin's finches aren't finches, African violets aren't violets, and electric eels aren't eels. Much worse, there are too many common names, and also not nearly enough. There are too many in that a single creature may bear many common names; for example, a single New World wild cat has at least 40 different English names (cougar, puma, catamount, panther, painter, mountain screamer, mountain lion, and many more). Those are just its English names, though; it also has names in French and Spanish and Portuguese, Nuu-chah-nulth and Q'echi' and Urarina, and dozens of other languages besides. This situation makes the keeping-track problem harder, although perhaps it isn't fatal to it. What *is* fatal to it is that in other ways there aren't enough names. Often, a single common name refers to many creatures: either because nonspecialists don't recognize the subtle traits that separate a group of similar species, or because the same name gets applied to different organisms by different people in different places. A *robin* is an entirely different bird in Europe and North America, for example, and the same confusion exists for *blackbirds* and *badgers;* a *fruit fly* can be any of several thousand species in at least two fly families, and a *daisy* can be almost anything. Worst of all, many species lack any common name at all (for example, the great majority of worms and insects). How are we to talk about these species at all?

Attempts to name organisms, and to list them systematically, are ancient. We have Babylonian clay tablets from 612 B.C. listing and naming about 200 species of medicinal plants. A Chinese text (*The Divine Farmer's Materia Medica*) listing 365 species was likely written in about A.D. 250, but codifies oral traditions as much as 3,000 years older. Aristotle and Theophrastus in ancient Greece (circa 300 B.C.) and Dioscorides and Pliny the Elder in ancient Rome (circa A.D. 50) named hundreds of animals and plants, and many of their names survive today.

The lists of known species that early scholars had to tackle were manageable, though. By the 1600s, the strain was beginning to show, as treatises began to appear covering thousands of species. Gaspard Bauhin's *Illustrated Exposition of Plants* (1623) covered 6,000 species, and John Ray's *Historia Plantarum* (1686) covered more than 18,000. These early modern works assigned Latin names that seem much like our modern practice, except that many of them are very long. Bauhin, for example, named one species of asphodel *Asphodelus foliis fistulosis* (or the asphodel with tubular leaves) and another *Asphodelus purpurascens foliis maculates* (roughly, the asphodel with leaves stained purple). These had nothing, though, on the name Peter Artedi bestowed (in 1738) on the English whiting (a fish; now *Merlangius merlangus*): *Gadus, dorso tripterygio, ore cirrato, longitudine ad latitudinem tripla, pinna ani prima officulorum trigiata.* Why such unwieldy names? Because the names were being asked to serve two functions at once: *naming* species and *describing* them (and the descriptions were expected to distinguish the named species from their relatives). The problem is that the more species were listed, the more cumbersome the descriptive names had to be. Perhaps worse, the discovery of new species would require adjustments to older names, so that each species' name continued to distinguish it from other similar ones.

Even in the seventeenth century, the names-as-descriptions system was creaking under the weight of known biodiversity; but it was obvious that it was only going to get worse. It was the great Swedish naturalist Carl Linnaeus who solved the problem. Linnaeus's innovation was to separate the two functions of naming: to make a species' name merely an identifying label, which can be used to look up the description (and anything else about it) in the literature. Actually, that wasn't quite what Linnaeus thought he was doing: he thought of himself as giving each species *two* names: a short label and a longer descriptive name (like the examples of Bauhin's above). But before long it became obvious that it was the short labels people would adopt as names. Those names were used as indexers to the descriptions, which were long and technical and which nobody was likely to memorize, trot out during a walk in the woods, or accommodate in their writing. Linnaeus's system of names-as-labels is the one we still use today.

Freed by Linnaeus from the yoke of description, Latin names could be short but still unique. Linnaeus's short names were what we now recognize as "binomials," with each species having a genus name and a species name of one word each. (A genus is just a group of similar, and as we now know, evolutionarily related species.) For example, *Homo sapiens,* our own species, belongs to the genus *Homo,* and within that genus we're the species *sapiens.* We're the only living member of *Homo,* but our extinct relatives include *Homo erectus, Homo neanderthalensis, Homo naledi,* and so on. Notice that among these names, *sapiens*—which means wise—and *erectus* are arguably descriptive, but they aren't what Linnaeus would have thought of as descriptive names, because they aren't precise enough to separate the named species from their relatives. *Neanderthalensis* and *naledi* aren't descriptive of their species at all: both refer to the locations of early fossil discoveries, in Germany's Neandertal region and in South Africa's Rising Star cave (*naledi* means "star" in the Sotho language). But names like *Homo sapiens* are short enough to remember and use in writing and speech, while being precise enough to mean the same thing to everyone everywhere.

Binomial Latin names are the most familiar, but it's worth noting that many organisms bear names that recognize finer divisions, below the species—and thus their names become trinomials. For example, the burrowing owl, *Athene cunicularia,* ranges through much of the

New World, and all individuals across its range bear that binomial name. However, burrowing owls in Florida differ in plumage from those across western North America, and this is recognized with "subspecies" names: *Athene cunicularia floridana* for the Florida birds, and *Athene cunicularia hypugaea* for western ones. (Another 20 subspecies have been named from the Caribbean and South America, although there are debates about the validity of some of them.) Subspecies tend to be recognized and named when variation within a species is structured geographically: when western individuals differ from eastern, island individuals from mainland, and so forth. Other patterns of variation are sometimes recognized with named "varieties," "subvarieties," and "forms," with obvious potential for complexity.

Subspecies and other trinomial names are used heavily in some groups of organisms (birds and butterflies, for example) and hardly at all in others, and one could be forgiven for considering them a messy complication. That they are—but they're a messy complication that's very important to what we think a "species" is. In Linnaeus's time, each species was held to have been specially created by God, and to have been unchanging since its creation. Subspecies names didn't make sense, in this context: if two distinct forms had been created, they'd be two species with separate names. But the revolution that followed Charles Darwin's publication, in 1859, of *On the Origin of Species* led scientists to think completely differently. Subspecies (and other sorts of variation below the species level) were now evidence for the mutability of species: the Florida burrowing owl could be understood as an isolated population in the process of evolving as a new species, distinct from its ancestor and from its sister Western burrowing owl. A flurry of trinomial naming in the latter half of the nineteenth century resulted as scientists rushed to document geographic variation as evidence for Darwin's new ideas. Etymologically, though, trinomials aren't much different from the binomials they supplement—and so, for the remainder of this book, we can afford to gloss over the distinction between a species name and a subspecies one.

By freeing names from the necessity of description, Linnaeus's invention of the Latin binomial opened the door to much more creativity in naming. Descriptive names had been heavily weighted to features such as morphology (the hollow-stemmed asphodel, *Asphodelus fistulosus,* with *fistula* Latin for a hollow reed) or color (the European fire ant, *Myrmica rubra,* which is reddish). Beginning with Linnaeus, names could refer to almost anything (and we'll see in the next chapter, "How Scientific Naming Works," that they do).

The key to the book you're holding now is that Linnaeus opened the door to eponymous naming. He took advantage of the open door, too, naming species after earlier botanists and zoologists (for instance, black-eyed Susan, *Rudbeckia,* after Olof Rudbeck), after his patrons (henna, *Lawsonia,* after Isaac Lawson, who bankrolled publication of Linnaeus's masterwork *Systema Naturae*), and even—in a way—after himself (twinflower, *Linnaea borealis*). By making names shorter, Linnaeus freed us to make *more* of them, and to make them more creatively. And while it wasn't Linnaeus's goal, that creativity gives us a window on the culture of science and the personality of scientists. Through the names we give species, we can see science as an indelibly human endeavor.

Linnaeus never knew just how big an advance his new naming system was, because he had no idea just how many species there were to be named. He thought there might be 10,000 plant species on Earth, but about 350,000 have been named so far—and his estimates of diversity for other groups were probably even further off. We now know that we need names in the thousands and tens of thousands and, eventually, in the millions. That need for names presents scientists with a chore, to be sure, but also with an opportunity.

2
How Scientific Naming Works

Species need names—but how do they acquire them? The short answer is that the person who discovers a new species has the privilege of assigning a name to it. Two parts of that short answer need a little further explanation, though. First, what does it mean to discover a new species? Second, how is a name actually assigned?

Discovering a new species sounds both simple and romantic. An intrepid explorer pushing through an uncharted rain forest swings a machete, and through a newly made gap in the lianas sees a stunning scarlet flower utterly unlike any species previously known to science. Our explorer seizes the flower (and the plant that produced it) and heads triumphantly home, to proclaim the news and bask in the resulting fame. And sometimes it really does happen that way; but it's usually a bit more complicated.

Although recognizing that a plant or animal specimen belongs to a species so far unknown to science sounds straightforward, it can actually be quite difficult. That's mostly a result of evolution's incredible ability to produce variations on a theme. That a million and a half species have been named so far probably makes it obvious that no individual human can possibly recognize them all. Many of us, of course, know a fair bit about particular groups—birds, perhaps, or ferns, or jewel beetles. Imagine that the jewel beetles are your group, and you've just picked up a small, slender, metallic beetle that you're quite sure belongs to the genus *Agrilus.* Is it an *Agrilus* species we already know, or a new one? There are over 3,000 species of *Agrilus* known so far, and some of them differ in ways that are quite subtle and require considerable technical knowledge to recognize. Your next move, then, is probably to ask an *Agrilus* expert. Such experts do, fortunately, exist, because a number of *Agrilus* species are economically important (among them *Agrilus planipennis,* the emerald ash borer, which is currently devastating ash trees across much of North America). Your *Agrilus* expert probably can't recognize all 3,000 species on sight either, but knows which books and journal articles to consult. It may take hours, or days, to work through the literature; but once you've finished, imagine that your beetle doesn't quite match the published description of any of the known *Agrilus.* Have you discovered a new species?

Well, maybe; but maybe not. In every species, individuals vary, and the degree of difference that should convince you that your *Agrilus* specimen represents a new species rather than an oddball may be far from obvious. Is your specimen just a largish *Agrilus abditus,* a smallish *Agrilus abductus,* a flattish *Agrilus abhayi,* a greenish *Agrilus absonus,* or a slightly aberrant *Agrilus aberrans?* It's not the observable differences that define a species anyway, or at least not directly. Instead, a species is (with complications that could fill several books) a set of individuals that have the potential to exchange genes by mating. Differences among individuals help us *recognize* barriers to gene flow, but they don't map 1:1 onto those barriers. Sometimes, an oddball is just an oddball: bigger or smaller or flatter or greener than its friends as a result of a chance combination of genes or the influence of the environment in which it lived. In a particular group, certain morphological characters may be more reliable indicators of species status, and experts usually know which characters they are—perhaps the numbers and locations of bristles have proven reliable in the past, but color hasn't. In insects, the shapes of the genitalia are very often the most reliable characters, so your unknown *Agrilus* specimen may have some uncomfortably close inspection in its future. It usually helps to have not just one specimen, but many, as species status is more likely when differences are discontinuous (that is, when there's variation within each of two species, but without overlap between them). Finally, in the last couple of decades, DNA sequence data have proven extremely useful in testing species status—even turning up new, distinct species in the absence of any physical differences at all.

No single indicator alone, though, is perfectly reliable; and the complexities of evolution are such that doubt often remains. For that reason, a claim that your *Agrilus* is a new species will always be your best guess; or, put more scientifically, a hypothesis that can be tested or disputed by other experts in the future. Sometimes the evidence for your new-species hypothesis will be very strong; sometimes it may be weaker. Sometimes, in fact, it may be *very* weak. Take, for example, a European species of freshwater mussel called the swan mussel, *Anodonta cygneus*. More than five hundred times, someone has formally described a "new" species of *Anodonta*, only to have their specimens turn out to belong to *A. cygneus* after all. (The five hundred names conferred on these illusory, not-actually-new species are now "synonyms" of *Anodonta cygneus*—about which, more below.) It turns out that freshwater mussels are notoriously variable, developing shells of different shapes on hard bottoms versus soft ones, in rapid flow versus slower, and so on; essentially, it's very easy to find an oddball mussel. There was, to make this worse, a certain nineteenth-century enthusiasm for naming new mussels based on very subtle distinctions—not just the real oddballs, but the only slightly eccentric. Mussel biologists are still clearing up the resulting mess, and (fortunately) it's since become conventional to hold new species hypotheses to a stricter standard.

All this complexity explains why new species aren't usually "discovered" in the field (with or without a swinging machete). Instead, most are discovered and described from specimens in the laboratory, or in a museum collection, often long after their collection from the wild. There, a biologist can compare specimens to those of previously described species, dissect tiny parts such as insect genitalia, extract and sequence DNA, and consult three centuries' worth of taxonomic literature. Museum collections are especially important here. In part, their role is to hold a field-collected specimen for later study—upon which it may prove to be a new species requiring description and naming. But studying a single specimen in isolation isn't likely to be informative, so it's even more critical that museums hold large collections, with many specimens of many different species (showing not just the diversity of species but variation within each). It's by reference to these large collections that a scientist can make the comparisons needed to achieve two important things. First, it's based on comparisons with collections that a species can be recognized as "new"—as truly distinct from all the world's previously named species. Second, comparisons with collections let a new species be placed among its relatives: perhaps it's a new member of a well-known genus, or perhaps it's something startlingly different that needs not just a new species name but a new genus name too (or even, sometimes, a new family or order or class). In a sense, you can think of each new species as being discovered *twice:* once by someone who collects it in the field, and then a second time by someone who (later) recognizes its novelty and decodes its relationships to more familiar species. The two discoverers may sometimes be the same person, but more often, they aren't.

What about your *Agrilus* beetle? If you and your *Agrilus* expert are convinced that it represents an undescribed species (or more accurately, if you're willing to go public with the hypothesis that it does) then it needs a name. That naming is up to you, except that it has to comply with a set of rules that keep naming from descending into chaos—something that separates scientific names from common ones. Species' common names appear and evolve in the same unruly fashion as do the rest of the nouns in our languages; but formal scientific names are different. Or at least, they are now. Early in the history of naming, it was a free-for-all, with scientists applying whatever name they wanted and changing earlier names willy-nilly. All this frantic naming activity threatened the very stability and precision that made scientific names useful, and so scientists developed formal systems for creating names and for deciding which name applies to each species. These systems evolved through the nineteenth and twentieth centuries and today are recorded in a set of rather legalistic Codes that govern the practice of naming. The Codes, I'll admit, don't make for riveting reading; but fortunately, we don't need to engage with them in detail. It will be enough for our purposes to see how the basic rules make it possible for everyone to agree on what to do in two important situations. When a species doesn't have a scientific name, they specify how a new one can be constructed and assigned. And when a species has been given two or more scientific names (remember the 500 names of *Anodonta cygneus*), they help us agree on which one to use.

You may have noticed that I referred to "Codes," in the plural. It would be nice if there were a single set of rules covering the naming of all organisms, but unfortunately, for largely historical reasons that isn't how it works. Instead, there are five separate Codes: one for animals, one for wild plants, algae, and fungi, one for cultivated plants, one for bacteria, and one for viruses. Of these, the first two are most relevant to this book, and although there are quite a few technical differences between them, they're nearly identical in spirit. Each Code provides a detailed and rather lengthy set of rules, but here are the basics:

- A new species or genus name appears when it's published in the literature, along with some supporting bits of science: notably, a description, and designation of a reference specimen (normally preserved for future study) known as a type specimen. "Published in the literature," by the way, is defined rather generously, such that names don't have to be published in a scientific journal. Almost any venue will do, as long as it's a printed or electronic document that's produced in multiple copies or is accessible from some outlet other than the original author or publisher. This matters because it means that amateurs can name species, and their names are just as valid as those of professional scientists. Finding published names from the 1800s (the heyday of amateur naming) can be something of an adventure, as species were named in popular nature books, travelogues, and even pamphlets with small press runs—and all this in the days before widespread indexing of publications.

- A new name can be coined in almost any way, as long as it meets a few simple criteria. Among them: it must be spelled with the modern Latin alphabet (the one we use for English), without special characters or marks such as accents or apostrophes (hyphens are allowed under certain conditions). The root of the name needn't be from the Latin language (and we have species names with etymologies traceable to hundreds of languages, or to none at all). However, once the root is decided, it's treated as if it were Latin, using some suffixes and Latin grammar that needn't concern us much. (This Latinization is why we commonly call scientific names "Latin

Northern gannet, *Morus bassanus*

names," even when their origin is from some other language.) Each species or genus name must be at least two letters long and reasonably pronounceable. Finally, a species name must be distinct from any pre-existing name in the same genus, and a genus name must be distinct from any pre-existing genus name governed by the same Code. That last rule removes much of the confusion that bedevils common names like "robin" and "badger" and "daisy." It ensures that there can be only one species of mulberry, *Morus,* with the species name *alba* (the white mulberry that's the preferred food for silkworms). Furthermore, no other plant genus can ever be named *Morus.* Mind you, gannets, under the zoological Code instead of the botanical one, are *Morus* as well—but there's little risk of mixing them up.

- When two or more names might be applied to a single species, the valid name is (almost always) the one that was published first. More recent names are "junior synonyms," and aren't used (although they remain in older literature and, absent careful reckoning, can be a source of confusion). The major exception to this "principle of priority" is that we have to start somewhere, so for convenience botanical naming is defined as starting with the first edition of Linnaeus's *Species Plantarum* (1753); zoological naming is defined as starting with the tenth edition of Linnaeus's *Systema Naturae* (1758). Older names are simply ignored.

Interestingly, the Codes don't have any legal force. Instead, the biological community (mostly) complies with them in part because nearly all agree that without some clear rules, we're lost; and in part because journals are unlikely to publish work that isn't in compliance.

There's one final detail that helps us keep track of names and their origins. Every scientific name has an "authority," which is the name of the person who did the original naming. So, for example, *Agrilus planipennis* has the authority "Fairmaire," because it was originally named (in 1888) by the French entomologist Léon Fairmaire. You'll sometimes see the name written including the authority—"*Agrilus planipennis* Fairmaire," or even "*Agrilus planipennis* Fairmaire (1888)." This is a shorthand for "*Agrilus planipennis*—you know, the one described by Fairmaire," which is useful because it lets us track down the original species description. Sometimes you'll see the authority name in parentheses—such as in the great horned owl, *Bubo virginianus* (Gmelin). This indicates that the species was first described as part of a different genus, and only later assigned to the genus in which it now sits. That can happen when a large genus is split up, when smaller genera are combined, or when we simply turn out to have been wrong about the closest relatives of a new species. Nomenclature can get messy! Finally, specifying the authority prevents confusion when the same binomial is applied to more than one species. This happens occasionally and accidentally, usually when a namer is unaware of the earlier use of an appealing name choice. It violates the Code, though, so the later name is rejected and must be replaced as soon as the situation comes to light.

You may have noticed that the Codes allow very wide latitude in how a new scientific name is constructed. That's what makes naming such an interesting and creative act. A species' scientific name can describe its appearance (the goldenrod *Solidago gigantea,* which is quite large as goldenrods go) or the sound it makes (the corncrake, *Crex crex*, which calls its own name). It can indicate where it lives (*Amolops hongkongensis*, a frog from Hong Kong) or what habitat it favors (the sergeant-major fish, *Abudefduf saxatilis*, with *saxatilis* meaning "living among rocks"). It can refer to mythology or religion (the baboon *Papio anubis*, named for the Egyptian god; or the cave-dwelling catfish *Satan eurystomus*). It can be a joke (the beetles *Agra vation* and *Ytu brutus*), or even an arbitrary combination of letters (the sponge *Hoplochalina agogo*). Or, finally, it can honor a person. That person might be the collector who first brought the species to scientific attention (the millipede *Geoballus caputalbus*, originally collected by George Ball and Donald Whitehead; *caputalbus* is simply Latin for "white head"). It might be the naming scientist's partner, friend, or relative (the goldenrod *Solidago brendae*, after the namer's wife, Brenda). It might be a benefactor (the lemur *Avahi cleesi*, after John Cleese, who donated to its conservation) or a celebrity (the spider *Aptostichus stephencolberti*). It might be a prominent naturalist (the tinamou *Nothura darwinii*); or occasionally, an obscure one (the snail genus *Spurlingia;* more about that snail in Chapter 7). The list of eponymous possibilities goes on.

How many scientific names are there, and what fraction of them are the eponymous ones that are the subject of this book? Neither question is easy to answer. A count of published names would be larger than the number of known species, because of the repeated-naming problem. *Anodonta cygneus*'s 500 names are an outlier, but it's not unusual for one species to have two or three, or half a dozen, synonyms in addition to the valid name established by the principle of priority. Those synonyms don't remain in use, but they were coined and thus have stories to tell. Nobody knows, though, just how large the total count of names might be, or how much larger it is than the count of valid names. There's no single global database of scientific names to consult, or at least, there isn't yet. If we take our best estimate for described species (1.5 million) and double it as a wild but conservative guess, we'd estimate that about three million names have been coined and assigned since Linnaeus codified the binomial system. Among them must be hundreds of thousands of eponymous names. One recent compilation by Lotte Burkhardt covered 14,000 eponyms just among the names of plant genera; and in the large genus *Aloe,* nearly a third of species names are eponymous. It would be the work of a lifetime to make similar estimates across Earth's stunning biodiversity; but it's clear what Linnaeus's invention of eponymous naming has given us. Hundreds of thousands of names tell stories—stories of the people honored in the naming, and stories of those who coined the names. This is a rich vein already, but it will only get richer, as the millions of undescribed species on Earth offer opportunities for more eponyms and more stories.

In the chapters to come, we'll explore just a few of the stories that eponyms tell. Here we go.

3

Forsythia, Magnolia, and Names Within Names

Early each spring, when the lawns and gardens of my hometown are still mostly gray and brown, and snowbanks linger in the shadow of slopes and buildings, the buds begin to break on a few of our earliest flowering trees and shrubs. I'm always eager for the splashes of color these give to a chilly spring day: the cheerful yellow of forsythias, and the elegant pinkish cream of magnolias. I've known these flowers, and known their names, for decades; but only recently have I realized that those names carry other names nested within them like etymological Russian dolls.

Forsythias and magnolias are two cases in which—somewhat unusually—the familiar common names are also, letter-for-letter, the Latin names of their genera. *Forsythia* is a small genus of about a dozen species, most of them native to east Asia. *Magnolia* is more diverse, with around 200 species native to east Asia and to the New World. Members of each genus are now grown in gardens around the temperate world, and when they're in bloom they're among our most instantly recognizable trees and shrubs. Less recognizable? The eponyms inside each name. If pressed, I suppose I might have guessed that the name *Forsythia* has a human "Forsyth" behind it, although I couldn't have taken that any further. The etymological roots of *Magnolia* I had no inkling of; but as it turns out, there was indeed a human "Magnol." As you might expect, there's a story behind each name.

Southern magnolia, *Magnolia grandiflora*

Forsythia was named, in 1804, by Martin Vahl. Vahl was a Norwegian botanist who became a student of Linnaeus, and published a number of compendia of plant names. As Vahl was working, the plant we now know as *Forsythia* came to the attention of European botanists. Another of Linnaeus's students, Carl Per Thunberg, assigned a Japanese species the name *Syringa suspense*. That would make it a lilac (genus *Syringa*), but Vahl, correctly, surmised that this wasn't a good identification, and proposed the new genus name *Forsythia* instead.

Although Vahl didn't explicitly say so, it's safe to assume that *Forsythia* honors William Forsyth, a Scottish botanist and horticulturalist. Forsyth was a founding member of the Royal Horticultural Society, the superintendent of two of the English Royal Gardens (at Kensington Palace and St. James's Palace), and a widely read expert on tree diseases and injuries. He was also, at the time of Vahl's naming, a controversial figure among botanists. Forsyth had invented what he called a "plaister"—a concoction of ashes, dung, urine, soapsuds, and other unpleasant ingredients that he claimed could be applied to a damaged tree to cure defects in its wood. This mattered because the

British navy was in desperate need of oak timbers for warships to fight in the French Revolutionary and Napoleonic wars. Forsyth was given a stipend of £1,500—about £130,000 in today's currency—from the British government to support his work on the plaister. This rather generous award was noted with derision by some of Forsyth's botanical rivals, with challenges, wagers, insults, and injured feelings ensuing on all sides. Vahl, in naming *Forsythia* in Forsyth's honor, may well have been picking a side: the pro-slathering-tree-trunks-with-dung side, if you will. It wasn't until after Forsyth's death, and after the naming of *Forsythia,* that the horticultural community came to a consensus. Forsyth's plaister was worthless; but *Forsythia* was, and is, beautiful.

Magnolia's story is very different. It was named in 1703 by Charles Plumier, a French botanist who made three botanical expeditions to French possessions in the West Indies. Plumier's *Magnolia* came from the island of Martinique. He gave it the full name *Magnolia amplissimo flore albo, fructu caeruleo,* or "the magnolia with large white flowers and blue fruit" (Plumier was working before Linnaeus's invention of the binomial; the species is known today, rather more succinctly, as *Magnolia dodecapetala*). The species description included an effusive dedication of the name, which began with reference to "the illustrious Pierre Magnol, counselor to the King, Professor of the Academy of Physicians and Professor at the botanical garden at Montpellier."[1] This makes Pierre Magnol sound a lot like Forsyth: a senior member of the botanical establishment, highly placed in society as a consequence of his royal appointment. But this telling of Magnol's story is misleading, and the real story is an important one that should resonate with us still today.

Pierre Magnol was born in 1638 in Montpellier, in southern France. Montpellier was a major educational and commercial center in Renaissance France, with a famous medical school. It also had France's first botanic garden, the Jardin Royale de Montpellier, which was dedicated to the teaching of medicine and pharmacology (in the sixteenth and seventeenth centuries, botany and medicine were so closely intertwined as to be practically a single subject). Montpellier was therefore the perfect place for Magnol, whose own interests were in botany and medicine, and by 1659 he had completed his medical training. He didn't practice, though, preferring to roam the countryside studying and collecting plants; his first major publication would be a catalogue of the flora of the Montpellier region. In 1668, two faculty chairs became vacant at the university, and Magnol was considered for the posts along with four other botanists. His reputation was outstanding and he outperformed the other candidates in an examination, and Magnol's name was submitted to the king for royal appointment. It was rejected, though; not because the king thought poorly of Magnol as a botanist, but because Magnol was a Protestant—a member of the minority Huguenot community.

Before Magnol's birth, France had suffered from a Catholic-Protestant civil war that ended in 1598 with King Henry IV's proclamation of the Edict of Nantes. The edict granted civil rights to France's Protestants, including the right to civil appointments such as the university chairs for which Magnol was competing. However, rights that exist in law aren't always respected in practice, and by the late seventeenth century the Protestant community in France was suffering considerable discrimination, both unofficial and official. King Louis XIV (Henry's grandson) was particularly hostile to Protestants, among other things denying them official appointments and forcing them to quarter abusive Royal Dragoons in their houses. Passing over Magnol was just one small piece of Louis's increasingly overt campaign to undermine the Edict of Nantes. In 1685, he simply revoked it, leaving Protestants like Magnol only three real choices: to live under active persecution, to leave France, or to convert to Catholicism. Hundreds of thousands left, but Magnol reluctantly converted. His new Catholic credentials, in 1687, at long last got him an official appointment: as demonstrator of plants, teaching botany to medical students. It wasn't a faculty chair, though; most likely, Magnol's new Catholic credentials were still a bit too suspect for a more prestigious appointment. Finally, in 1694, Magnol was appointed to a newly vacant chair at the Jardin Royale de Montpellier, as a professor of medicine and Director of the Garden. Magnol was 56 years old, and for more than three decades he'd been denied similar appointments despite his reputation as one of France's most able botanists.

Magnol's most enduring contribution to botany lay in his *Prodromus,* published in 1689. This was an attempt to provide a listing and general classification of the world's plants, and it was the first such attempt to arrange plants into families (following previous practice for animals). Earlier treatments of plants had, most often, simply listed them alphabetically, but with his plant families, Magnol tried to recognize what seemed like natural groupings of plants with similar characteristics. There would be a lot of rival classification systems over the next centuries. Magnol's was not only the first, but one of the better early attempts. That's because he resisted the temptation to give priority to a single kind of trait (as Linnaeus would do 50 years later, for example, with a system based entirely on counts of floral parts). Instead, Magnol wrote that "there is a certain likeness and affinity in many plants which does not rest upon parts taken separately but in the total composition, which strikes the sense but cannot be expressed in words."[2] This realization started us down the road to our modern understanding of plant relationships: for example, petunias, tomatoes, and potatoes are together in the family Solanaceae; daffodils and garlic in the Amaryllidaceae; and roses, raspberries, and apples in the Rosaceae. This is handy for organizing and learning plant diversity, of course, and that's what Magnol was intending with his system of families—but ultimately it's far more important than that. Magnol didn't know it (and may well have had religious objections if he did), but his organizational system was one of our first steps toward realizing that all plants, and all life on Earth, have a common evolutionary origin and a shared evolutionary history. It's that evolutionary history that accounts for our ability to organize them as he did—we group daffodils with garlic because they share features, and they share features because they're close evolutionary kin. This fact is the foundation of all modern biology, and every spring the blossoms on our *Magnolia* trees celebrate Magnol's contribution—remembered today by very few, but none the less important for that.

How severely was Magnol's contribution limited by the religious discrimination he endured? It's hard to know. Even without an official position, he managed to build a considerable reputation, hosting a visit from the great English botanist John Ray and having his flora of the Montpellier region praised by Linnaeus. It probably helped that he came from a wealthy family of apothecaries and so had the means to pursue botany even when official positions were denied to him. But how much more might he have done? And how many other Protestants, with fewer family advantages, were sidelined more completely? There's an important lesson here for us still today, and it's rooted in an irony in Magnol's eventual appointment to teach medicine in Montpellier. Renaissance medicine, and especially the curriculum of the medical school at Montpellier, owed an enormous amount to knowledge generated by Arab physicians and scientists. Science in general, and medicine in particular, had thrived for centuries in the Islamic world while Europe endured its medieval dark ages. The obvious lesson—that contributions to human progress transcend boundaries of nationality, race, and religion—was lost on seventeenth-century France. It's still lost on many today. We've made a lot of progress in opening up science and our other human institutions to women, to people of color, to those of all sexualities, and so on; but there's a lot more still to be done. Xenophobia and intolerance remain alive and well, and are even enjoying something of a moment in the sun as politics in many countries tilt toward right-wing demagoguery. Perhaps, in this context, it would be useful for us to see *Magnolia's* blossoms each spring as reminders of an intolerant past, and as a suggestion of hope for a more inclusive future.

There are stories, then, in *Forsythia* and in *Magnolia.* Those stories involve botany, of course, but also history, and personality, and conflict, and (at least for Magnol) achievement in the face of adversity. Neither Forsyth's story nor Magnol's is much told today, but each Latin name anchors the story and flags it for the curious to find. Across the diversity of life, thousands of other eponymous names do the same.

4

Gary Larson's Louse

Some of Earth's creatures are majestic, like California redwoods or the harpy eagle. Some are achingly beautiful, like birds of paradise or the showy lady's-slipper orchid. Others, like the great white shark, are terrifying. Still others, like the polar bear, are all of those things at once. Having a creature like one of these named after you must surely be an honor, and a thrill.

Gary Larson got a louse.

Gary Larson is the cartoonist behind *The Far Side,* which ran in newspapers from 1980 to 1995. It's really not possible to explain *The Far Side* to those who have never read it—"quirky" would be a major understatement—but nature and the scientists who study it were very frequent topics. There were cartoons about slug dinner parties, about spiders spinning webs across playground slides ("If we pull this off, we'll eat like kings"), about porcupines with punk hairdos, and about amoeba conventions complete with tiny nametags. Larson's cartoons were often absurd, but the absurdity was rooted in a fascination with the oddities of nature—the very oddities that draw a lot of biologists to their studies. As a result, *The Far Side* was beloved by biologists, and photocopied *Far Side* cartoons still decorate laboratory doors in universities, museums, and research institutes everywhere. Sooner or later, it was inevitable that someone would name a species in Larson's honor. The first someone to do so was Dale

Clayton, an entomologist studying the lice that feed on the feathers of birds.

We're all familiar (some of us distressingly so) with the lice that feed on humans. Three species afflict us: the head louse, the body louse, and the pubic louse. These species are, however, only the tip of the liceberg. Worldwide, we know of about 5,000 species of lice, and there are probably thousands more yet to be described. Lice are diverse, we think, because they tend to be strict diet specialists: the human head louse won't infest a rhesus monkey, and a rhesus louse won't broaden its horizons to feed on a human head. This fussiness has driven an evolutionary radiation of lice alongside their bird and mammal hosts, and that radiation includes several diverse lineages that feed by chewing on bird feathers. One feather, you might think, is a lot like another; but in fact many feather lice are specific to single bird species, or to a few.

Dale Clayton, in research for his Master of Science degree, studied a genus of lice specializing on owl feathers: *Strigiphilus* (appropriately enough, the genus name is based on Latin meaning "owl lover"). In a paper published in 1990, Clayton described three new species of *Strigiphilus,* naming one for his graduate adviser (*Strigiphilus schemskei*), one for a scientist colleague (*Strigiphilus petersoni*), and one for Larson (*Strigiphilus garylarsoni*). Clayton was (and still is) a *Far Side* fan, and speaks of his appreciation for Larson in two ways: first, his insightful understanding of how nature works; and second, the role of *The Far Side* in spreading interest in nature—because "there's no better teacher than humor."[1]

S. garylarsoni is a tiny insect, at most 2 millimeters long, known only from a small African owl (the southern white-faced owl). It's distinguished from its close relatives by some subtle characters that could matter only to louse taxonomists and to other lice: the length of some

Gary Larson's louse, *Strigiphilus garylarsoni*

hairs on the head, and the shape of one piece of the male genitalia. It isn't brightly colored or elegantly sculptured, it doesn't sing a beautiful song, and it isn't the linchpin holding together an ecosystem. But it belongs to Gary Larson, with Clayton's dedication "in appreciation of the unique light he has shed on the workings of nature."[2]

You might, of course, wonder how the recipient of such an unusual tribute might feel about it. For some, immortalization in the name of a tiny obscure parasite might not seem like something to be celebrated. Clayton wondered too, and before bestowing the name he wrote to Larson explaining his intention, and asking if Larson approved of "this somewhat dubious honor." He did. In fact, Larson included Clayton's letter in his book *The Prehistory of The Far Side* in 1989, along with a photograph of *S. garylarsoni* and the remarks "I considered this an extreme honor. Besides, I knew no one was going to write and ask to name a new species of swan after me. You have to grab these opportunities when they come along."[3] Larson also had a line drawing of *S. garylarsoni* reproduced, in ranks and files 500 lice strong, as the endpapers of his book. With *Prehistory* having sold more than two million copies, Clayton is both bemused and (understandably) a bit proud that this drawing might be the most-reproduced scientific image of all time. For nearly 30 years since their first communication, Clayton and Larson have kept in touch, exchanging Christmas cards and occasional dinners. Larson even supplied a jacket blurb for Clayton's most recent book, a technical work on the coevolution of parasites and their hosts—making him surely the only evolutionary biologist ever to have his scientific work blurbed by a cartoonist.

It's a happy story, then, the tale of *Strigiphilus garylarsoni,* even if it doesn't involve a swan. Which is good, because Larson's other brush with eponymy proved more fleeting. In 1990, Kurt Johnson published a revision of the Neotropical butterfly genus *Calycopis* (and a few of its relatives). The result was a dramatic splitting up of the group and a profusion of new names, with 235 species names spread across 20 genera. Among Johnson's new genera was *Serratoterga,* and among his new species was Larson's hairstreak, *Serratoterga larsoni.* So Gary Larson got a beautiful butterfly to go with his louse. Or at least, he did for a while. Fourteen years later another entomologist, Robert Robbins, published his opinion that *Serratoterga larsoni* isn't a distinct species at all. Instead, Robbins argued that the butterflies Johnson called *S. larsoni* belong to the very familiar species *Calycopis pisis*—as do those to which Johnson gave several other new "species" names. Johnson was what taxonomists call a "splitter"—he was inclined to recognize even minor variants as new species (or even as distinct genera). Robbins was a "lumper," who argued that all species include individuals that vary in genetics, morphology, and behavior. Where a splitter sees dozens of narrowly circumscribed species with variation between them, a lumper will see a single species with variation within it, and it's an argument that's gone on as long as scientists have described species. Entomological consensus now sides with Robbins's lumping: Johnson's naming of *S. larsoni* wasn't necessary because the butterflies he applied the name to were just slightly different individuals of *Calycopis pisis,* rather than representing a newly discovered, distinct species. Larson's supposed hairstreak already had a name—and it had had that name for more than a hundred years.

The name *Serratoterga larsoni,* in technical terms, is a "junior synonym" of *Calycopis pisis,* and is no longer in use. The science of taxonomy is littered with ghost names like this, because disagreements between taxonomists—between lumpers and splitters, for example—aren't unusual. There are even times when opinion about species boundaries ebbs and flows across a whole discipline. Consider, for example, what happened to the consensus number of bird species in the 1920s and 1930s. At the beginning of this period, most ornithologists recognized around 19,000 species of birds worldwide; but by its end, the list had been trimmed below 9,000. Some names were discarded, as *Serratoterga larsoni* was, while others that recognized geographic variation within species were transformed into subspecies designations. Of course, names can come and go, but all that variation is still there—so in 2016, a group of ornithological splitters led by George Barrowclough took a shot at turning the tide with a paper arguing that there are, in fact, closer to 18,000 bird species than 9,000. If consensus settles once again behind the larger number, a lot of disused species names will be dusted off again. It's fortunate, we can agree, that that isn't going to happen to the swan mussel *Anodonta cygneus* and its 500 splitter-derived synonyms. It isn't likely to happen to *Serratoterga larsoni,* either.

Serratoterga larsoni, then, is something of a lost name. It shows Johnson's *intent* to honor Gary Larson, but it's hard to see it as having achieved that honor. If Larson is to have a butterfly, another entomologist will have to bestow his name on another undescribed species; and this time, a real one. Gary Larson's louse, in contrast, remains *Strigiphilus garylarsoni,* honoring the cartoonist with every specimen collected, identified, pictured, or written about.

Gary Larson isn't, of course, the only person to have received the rather strange tribute of an eponymously named but inglorious species. Swans, butterflies, birds of prey, and orchids all need names (and among those, at least butterflies and orchids include many thousands of species that haven't yet received them)—but they're outnumbered by far by species that, like *Strigiphilus garylarsoni,* may be beautiful only in the eye of certain beholders. The planet teems with dull brownish beetles, tiny wasps, near-microscopic nematode worms—and mites. *So many* mites: hundreds of thousands of species at least, perhaps a million or more. They're found everywhere—in soils, on plants, in rivers and streams, even living among your eyelashes—but they're mostly dust-small and beloved only by acarologists (the scientists who study them). One of these mites belongs to Neil Shubin.

Neil Shubin is an evolutionary biologist and paleontologist best known for two things: (co)discovering the fossil *Tiktaalik roseae,* and hosting the PBS documentary series *Your Inner Fish* (based on his earlier book of the same name). *Tiktaalik* is a lobe-finned fish from the late Devonian, 375 million years ago, with features associated with the evolutionary transition from fishes to the first amphibians. Its discovery in 2004 made a big media splash. *Your Inner Fish* tells the story of a similar but more extended transition: that from fishes (and even deeper ancestors) to the human body that's reading this book right now. Shubin has thus made contributions both to the practice of science, as a paleontologist, and to the communication of science, as an author and television host. Both contributions caught the eye of Ray Fisher, a Ph.D. student studying a group of North American river mites in the genus *Torrenticola*. These mites have tiny larvae that are parasitic on midges, and almost-as-tiny adults (less than 1 millimeter long) that hunt for prey in sandy sediments on the bottoms of fast-flowing streams. When Fisher published his research in 2017, he described and named 66 new species of *Torrenticola,* including Neil Shubin's mite: *Torrenticola shubini.* He explained the name as being "in honor of author and paleontologist Neil Shubin for his efforts to popularize stories of human evolution with his book (2009) and TV series (2014), *Your Inner Fish.* As with many of the species that Shubin studies (e.g., *Tiktaalik roseae* . . .), *Torrenticola shubini* may represent a key evolutionary transition."[4]

Unlike Clayton, who sought permission from Gary Larson before naming *Strigiphilus garylarsoni,* Fisher simply published the name and then sent Shubin a copy of the paper—when *Torrenticola shubini* was already a mite accompli. There's some risk in naming that way, I suppose; but Shubin was very pleased with his eponymous mite. He explains it this way: "It's just a nondescript little mite, but it's my mite. It will probably outlive me, unless maybe the genus gets revised [note: that's exactly what happened to *Serratoterga larsoni*]. It's a cool little honor; it's in the literature, it has a life of its own. [It's] a lovely thing. The honor is the same, whether it's a new species of hominid, or a louse, or a mite—someone thought of you and your contributions as worthy of recognition."[5] There are more formal honors that can be bestowed on a cartoonist, or on a biologist; most of them come with more public visibility than an eponymous name in a scientific paper. Both Larson and Shubin have won some of them—Larson a Reuben Award from the National Cartoonists Society, for example, and Shubin a Communication Award from the National Academy of Sciences. There are Pulitzers and Nobels, too (although neither Larson nor Shubin has yet gotten that phone call). But at least among biologists (and

I'll count Larson a member of the fold), an eponymous name is something to cherish. In the words of Peter Collinson, the eighteenth-century botanist for whom Linnaeus named the plant genus *Collinsonia,* to be remembered in an eponymous Latin name is to be given "a species of eternity . . . as long as men and books endure."[6]

So perhaps it seems the strangest kind of tribute, to have your name attached for all posterity to a mite or a louse, but it's a sincere tribute, and at least for recipients who understand, it's a welcome one. There are a lot of mites and lice out there to be named, and a lot of other inglorious species too. That's not a bad thing. After all, we can't all have polar bears and birds of paradise, but there is eponymic hope for everyone nonetheless.

5

Maria Sibylla Merian and the Metamorphosis
of Natural History

The Argentine giant tegu lizard, *Salvator merianae*. Merian's spot-edged white butterfly, *Catasticta sibyllae*. Merian's sphinx moth, long-jawed orbweaver, giant stink-bug, and orchid bee: *Erinnyis merianae, Metellina merianae, Plisthenes merianae,* and *Eulaema meriana.* The bulbil bugle-lily, *Watsonia meriana.* Merian's dwarf morning-glories, *Meriana* spp., and Jamaica roses, *Meriania* spp. From a South American lizard to a common European spider to a spectacular bugle-lily of the South African fynbos, these species have one thing in common: their Latin names celebrate one of the most important, and most fascinating, women in the history of science. But they celebrate her in different ways, recognizing different facets of her interests, her accomplishments, and her importance to science. Their variety is, in a way, more than the sum of their parts.

Maria Sibylla Merian was born in Frankfurt in 1647. Life took her from printing shop to ascetic commune to high society, from Germany to the Netherlands to Suriname and back again. Combining a keen eye for botanical and entomological observations with outstanding skill as an artist, she published revolutionary books on the development, metamorphosis, and natural history of butterflies, moths, and other insects. She would have fame in her lifetime, although after

Maria Sibylla Merian in
1700 (Copperplate by
Jacobus Houbraken, from
Das Insektenbuch;
public domain)

her death she would be repudiated, then largely forgotten—and finally rediscovered.

In some ways, in her career path as a naturalist and an artist, Merian had every advantage. She lived at a time when curiosity cabinets were bulging with weird and beautiful specimens returned from burgeoning overseas exploration and trade. Her father was a publisher and engraver whose press issued lavishly illustrated works of natural history and geography. When he died (she was three), he left her a substantial fortune. Her stepfather and her husband were both artists, and the art of the day leaned heavily on depictions of nature and of still-life arrangements rich with flowers, insects, and natural curiosities. In other ways, of course, the stars were firmly aligned against her: she was a woman, and it was the seventeenth century. At that time, in at least some corners, a woman's interest in nature was seen as at best eccentric and at worst redolent of witchcraft.

Merian was always obsessed with plants and insects. As a girl, she would collect flowers and insects for her stepfather to paint, and she would help with engraving and illustration. According to one story, she stole a tulip from a neighbor's garden to paint, but he admired her work enough to forgive her and ask to have the painting. Later in life, Merian would date the origin of her serious interest in natural history to 1660, when she was 13 and began observing and sketching the development of silkworms. In 1679 she published *The Wonderful Transformation of Caterpillars and Their Remarkable Diet of Flowers,* a masterful book on caterpillars, butterflies, and moths.

The Wonderful Transformation made it obvious that Merian was a superb artist, but perhaps even more a scientist pushing at the boundaries of her field. She differed from other contemporary natural historians in her emphasis on understanding insect life history and development, and in connecting observations of eggs, larvae, pupae, adults, and host plants. Other scientists still had a static, specimen-centered viewpoint. For example, Thomas Moffet's *Theater of Insects* (1634) was an important reference work in Merian's day. This book illustrated caterpillars, pupae, and adult moths and butterflies—but in separate chapters for each life stage, and without clearly connecting which caterpillar went with which adult. This organization wasn't remarkable at all, since most scholars of the day didn't understand the cycle of insect life history from egg to adult and back to egg. Many, in fact, still believed in spontaneous generation. Daniel Sennet's famous *Thirteen Books of Natural Philosophy* (1660) assured the reader that although butterflies develop from caterpillars, "experience shows that such Worms and Caterpillars are bred out of Dew and Rains falling down upon Plants." For naturalists who failed to recognize the continuity of development from egg to larva to adult to egg, there was no particular need to organize studies of insects around the relationships between these stages.

Merian's work upended this attitude. She worked hard to collect eggs or caterpillars, to rear them through development on the plants they required for food, and to illustrate all stages, and the food plant, together. Many of her paintings even included the parasitoid wasps and flies that sometimes emerged from a pupa in place of the adult butterfly she was expecting. These mystified her; and while she argued strenuously against spontaneous generation for caterpillars and butterflies, until late in life she didn't rule it out as a possibility for parasitoids. So Merian's thinking represented a major advance, without being entirely correct. Merian's work didn't singlehandedly refute the doctrine of spontaneous generation, but her meticulous observations went along with experiments by contemporaries like Jan Swammerdam and Francisco Redi in building the case against it. It would take another 150 years for that case to become airtight, and for Merian's approach to insect life histories to be universally accepted and therefore unsurprising. All this is pretty typical of the progress of science, with steps forward accompanied, and sometimes clouded, by false starts and mistakes; and with advances building on the cumulative work of many scientists, not (or at least rarely) on the overthrow of a field by one brilliant loner.

Merian's reputation as an artist and naturalist was well established by the 1680s, and in the middle of that decade, she made the first of several reinventions of her life. She left her husband, Jacob Graff, and traveled to the Netherlands to join a religious (Labadist) commune at Wieuwerd. When her husband followed her there, she refused to see him. The Labadists had established colonies in the Americas, including Suriname, and in the commune Merian first encountered the tropical butterflies that would later become an obsession for her. When the commune disbanded in 1691, rather than return to her husband and her roots she moved to Amsterdam. There she established herself as an independent and well-respected member of society. She had a comfortable house, access to a steady stream of specimens returning with overseas trade, rich and important buyers for her work, and a community of artists and scholars to which she belonged. But it wasn't enough. She grew dissatisfied with the way other collectors and naturalists divorced organisms from their ecological context, growing tropical plants in glasshouses, dissecting dead specimens, and painting pinned butterflies and stuffed birds. Merian yearned to do for New World butterflies and moths what she had done for the more familiar European ones: watch and record living insects feeding, growing, and developing in nature. And so, in 1699, she reinvented herself again: she left Amsterdam for the Dutch colony of Suriname, to study the insects and other fauna of the South American rain forest. She took her 21-year-old daughter Dorthea with her, but left everything else behind.

Merian's voyage to Suriname was an extraordinary undertaking. She left Europe 69 years before Captain Cook first sailed the Pacific; 100 years before Humbolt's expedition to Central and South America; and 132 years before Darwin's famous voyage on the *Beagle*. In the 1690s, it was highly unusual for a European to go to Suriname for any reason other than to traffic in sugar and slaves. For a woman on her own, it was unheard of. Other seventeenth-century artists and naturalists traveled with funding and protection from a monarch or a trading company, but Merian funded her own expedition by selling 255 of her flower and insect paintings, and by promising to supply specimens to European collectors. She risked shipwreck, pirates, and the typically horrid and inadequate shipboard diet to make the voyage. But arriving wasn't a lot safer: Indigenous communities and escaped slaves were in intermittent revolt, the French were threatening invasion, and the rain forest teemed with venomous snakes, parasites, and mosquitoes carrying malaria and yellow fever. She began her stay in Paramaribo, then a town of fewer than 1,000 European colonists— many of them convicts pressed into service as soldiers, or sailors waiting for a ship in need of hands. It was not a place she fit in: she wrote that the colonists would "jeer at [her] that [she was] looking for other things than sugar in the country."[1] From Paramaribo she worked her way deeper into the jungles, collecting and painting as she went.

During her time in Suriname, Merian collected, studied, and illustrated not just butterflies but beetles, flowers, toads, snakes, spiders, birds, and much more. Sometimes she hacked paths into the famously impenetrable rain forest (or had servants, or slaves, hack them for her); sometimes she felled trees to collect caterpillars from the canopy. She wasn't the first European to describe New World rain forests (in 1648, Willem Piso had published a book about his travels in Brazil), but she was probably the first to see what their canopies held, and to describe how greatly that differed from the understory closer at hand. She was certainly the first to see the rain forest as an ecological web. She filled crate after crate with specimens, and notebook after notebook with observations and paintings. Her art became even more ecological than it had been in Europe: less ornamental and less sentimental, more disordered and richer in movement and evidence of nature red in tooth, claw, and insect mandible.

After two years, though, declining health forced Merian to cut short her expedition. It's likely that she had contracted malaria, although there were lots of equally unpleasant possibilities. She returned to Amsterdam, and seems to have picked up in social position more or less where she left off: about as much of a celebrity in scholarly circles as a woman could be at the time. She worked feverishly to complete what became her masterwork: the *Metamorphosis Insectorum Surinamensium* (*Metamorphosis of the Insects of Suriname*). This had

60 full-page plates and another 60 pages of Merian's observations of plants, animals, and Surinamese society. She offered the book by subscription, and a hand-colored printing of the first edition cost 45 florins—enough, as an alternative, to buy 1,300 pints of beer. In England, an advertisement soliciting subscribers referred to her as "that Curious Person, Madam Maria Sibylla Merian." Curious she was, in both senses of the word; and there were enough curious subscribers to support publication of the book in 1705 (and several later editions).

Merian died in 1717, just a few months short of 70 years old. Her work remained hugely influential through the 1700s, with her books widely read, cited, and admired. Linnaeus referred to her illustrations at least 100 times, and at least several times used her illustrations to describe species that he hadn't seen in any other way. (Linnaeus named two species for Merian: a moth, *Phalaena merianella,* and a butterfly, *Papilio sibilla,* but unfortunately neither name remains valid today.) By the mid-1800s, though, her reputation had become checkered at best. While some great naturalists still praised her books (including Henry Walter Bates, Louis Agassiz, and Alfred Russel Wallace), others criticized them. In 1834, for example, Lansdown Guilding published a detailed critique of the *Metamorphosis,* referring to the illustrations as careless, worthless, or even "vile," and accusing her of mistakes that should have been obvious to "every boy entomologist." Mind you, Guilding had never been to Suriname, and he was working from a poorly colored edition published well after Merian's death and including plates she didn't actually draw. None of that bothered him. James Duncan's *Naturalists' Library* (1841) suggested, quite incorrectly, that her illustrations were "to a considerable extent fabulous" (meaning by "fabulous" fictional, in the sense of a fable; if Duncan had meant today's sense of "fabulous" meaning terrific, he'd have been right). Another naturalist, William MacLeay, was particularly skeptical of one plate in the *Metamorphosis,* showing a tarantula poised to feed on a hummingbird. Surely, he argued, nobody seriously believed that tarantulas hunted in trees or that they captured and ate birds! Forty years later, and 170 years after Merian's original observation, the famous explorer and naturalist Henry Walter Bates confirmed that she was right and MacLeay was wrong. Merian's decline in reputation seems to have had little to do with errors (she made a few, but so did everyone else), and more to do with a Victorian attitude that all previous work was untrustworthy. Still, at least in the nineteenth century she was remembered; for much of the twentieth, she was completely forgotten.

Merian's work began a well-deserved renaissance in the 1970s, when the library of the Academy of Sciences of the USSR began issuing facsimiles of her insect books. Museums exhibited her work and told her story, and she appeared on stamps and a German 500-mark bill. She's still not a household name like Darwin or Linnaeus, but increasingly at least entomologists are aware of her remarkable accomplishments. We owe her a great debt, not only for having revealed many of the secrets of insect metamorphosis, but also for helping work a metamorphosis of a different sort: a reinvention of the way naturalists thought about plants and animals. Previous illustrations had been stylized, tidied up, and often paid more attention to ornament than to verisimilitude. Merian pioneered an ecological approach to natural history. Her work showed leaves and flowers with damage from insect attack, emphasizing connections between caterpillars, food plants, and enemies and the tangled complexity of nature. She is rarely credited with this advance. The biogeographer Alexander von Humboldt is more often called the first modern ecological thinker, but he didn't begin publishing until the 1790s. There is no question, though, of Merian's influence: natural history illustration, and natural history itself, were never the same after her.

What of Maria Sibylla Merian's eponymous species? These represent recognition of her importance, and the debt we therefore owe her, by scientists who followed—those who described species based on her illustrations, and those who found in their own name-thirsty collections ways to honor her contributions to art, to botany, to entomology, and to zoology. Those who named plant species for her—the New World *Meriania* and *Meriana*, and the South African *Watsonia meriana*—seem to have been emphasizing the beauty of her botanical paintings. Indeed, her first book (*Das Blumenbuch*, or *The Book of Flowers*) was a showcase of such beauty. But as beautiful as Merian's eponymous flowers are, they may sell her a little short. More appropriate tributes lie in her New World butterflies (*Catasticta sibyllae* and *Erinnyis merianae*); these insects were her lifelong passion, and her greatest scientific contributions were in entomology, not botany. *Catasticta sibyllae* is a particularly handsome butterfly, and its namers recognized Merian's role both as an artist and as a scientific researcher: "her research, in the form of numerous paintings . . . served as a foundation for the scientific study of insects."[2] I'm intrigued, too, by the naming of *Metellina merianae*, a common European spider that

The Argentine giant tegu, *Salvator merianae*

would surely have watched her paint. The species was named in 1763 by Giovanni Scopoli, with no explanation offered. The very lack of explanation, though, lets me imagine Scopoli smiling at the notion of Merian's spider watching Merian with the same curiosity as Merian watched the rest of the natural world.

On the whole, I suspect Merian would have been very pleased with the variety of species that now bear her name. But of all her creatures, perhaps the one that pleases *me* the most is *Salvator merianae,* the Argentine giant tegu. Why? Because it reminds us that although her passion might have been butterflies, her curiosity wasn't limited to them. Merian cast her keen eye on the full complexity of nature's web. The tegu she painted, on a cassava branch, is probably hunting the white peacock butterfly caterpillars feeding there, wrapping three levels of the food chain into one painting. *Salvator merianae* was named in 1839, more than 130 years after Merian published her illustration and well into the period during which some naturalists had become skeptical of her work. Nonetheless, the three French naturalists who gave the tegu Merian's name had no difficulty recognizing the species from her painting—on which they based their description. *Salvator merianae,* then, nicely sums up both the breadth of Merian's thirst for knowledge and her long-lasting influence on the way we think about nature. Nobody has ever deserved immortalization more.

6

David Bowie's Spider, Beyoncé's Fly, and Frank Zappa's Jellyfish

Scientists, in much of the popular imagination, are a little less than hip. Nerds, not to put too fine a point on it, with our scotch-tape-repaired glasses, our ink-stained-pocket lab coats, and our ivory-tower removal from the real world. And if this stereotype gets applied with particular vigor to any group of scientists, it would have to be taxonomists. Taxonomists are fussy old men (in the stereotype, always men) in stuffy rooms in the basements of museums, poring over drawers of dusty specimens and squinting with crossed eyes to find just the right trivial character to distinguish two species. The last thing on the mind of a scientist, and especially a taxonomist, is pop culture—right?

Well, not so fast. Actually, scientists are just people like everybody else, which means that among us you'll find nerds and bodybuilders, sippers of claret and swillers of Bud Light, opera buffs and Justin Bieber fans. It's true that most eponymous namings recognize historical fig-ures, other scientists, and the like, but you might be surprised to learn of the spider named for David Bowie, the horsefly named for Beyoncé, and the jellyfish named for Frank Zappa. These are just a few of the many species named for musicians, actors, and other pop-culture celebrities. Their existence casts taxonomy in a different light, establish-ing that no scientific field is immune from the inclination to play.

David Bowie's spider, *Heteropoda davidbowie*

David Bowie's spider, *Heteropoda davidbowie,* is a great example. It was discovered in Malaysia and named by Peter Jäger in 2008, but it had its 15 minutes of fame in the winter of 2016, when Bowie died at the age of 69 and every detail of his long life and career was breathlessly retold in the media. David Bowie, of course, was pop-music royalty. He first came to prominence in 1969, when the song "Space Oddity" was released just five days before the Apollo moon launch. Bowie's music was inescapable in the 1970s and early 1980s, and he released a final album, *Blackstar,* two days before he died.

Why a spider? Over his nearly 50-year career, Bowie reinvented himself constantly, adopting a series of musical styles and stage personas. Among these personas was Bowie's early-1970s alter ego Ziggy Stardust, who roamed the stage thin, long-legged, and orange-haired—and backed up by his band, the Spiders from Mars. *Heteropoda davidbowie,* quite appropriately, is itself thin, long-legged, and orange-haired. The best celebrity namings, I'd argue, are like this one: appropriate to the species, recognizing some feature of the organism or some other clever connection between species and eponym.

David Bowie and Beyoncé never shared a stage, but they share the distinction of having cleverly eponymous species names in their honor. Beyoncé, the American R&B and pop singer who's been as inescapable since the turn of the twenty-first century as Bowie was in his own heyday, is honored in the name of a horsefly. It's one of five new species added to the Australian genus *Scaptia* in 2011 by Bryan Lessard and David Yeates. Only three specimens have ever been collected, and they sat unidentified in museum collections for decades before being recognized as distinct from their relatives, and new to science. *Scaptia beyonceae* is distinguished from other *Scaptia* species by its "conspicuous golden tomentum on tergite four onwards"—or less technically, by its rounded and golden derrière.[1]

In explaining the name's etymology, Lessard and Yeates said only that the "specific epithet is in honor of the performer Beyoncé."[2] News coverage, though, didn't miss the implied reference to Beyoncé's booty; she is, after all, well known for favoring golden couture that flaunts curves both fore and aft. Does this push at the boundaries of good taste? Perhaps; scientists have been known to push at those boundaries just like everyone else. But then, in the 2001 single "Bootylicious," as lead singer of Destiny's Child, Beyoncé sang about her derrière with considerable enthusiasm. *Scaptia beyonceae* and its description, then, are presumably in agreement with the artist herself.

Of course, celebrity namings aren't limited to musicians. Scientists have interests that span our world's culture, both highbrow and low—as we have already seen in the story of the louse named for cartoonist Gary Larson, *Strigiphilus garylarsoni*. The television comedians Jon Stewart and Stephen Colbert have a wasp and a spider, respectively (*Aleiodes stewarti* and *Aptostichus stephencolberti*). Athletes have species, too—for example, the wasp *Diolcogaster ichiroi,* for

Ichiro Suzuki, who holds major league baseball's all-time record for hits in a season (262). The fantasy novelist Terry Pratchett has a fossil sea turtle (*Psephophorus terrypratchetti*), which makes sense if you know that Pratchett's novels are set on the Discworld, a flat planet resting on four giant elephants that stand in turn on a giant turtle swimming through interstellar space. Equally apropos, Herman Melville, who wrote about the great white whale Moby Dick, has a fossil sperm whale (*Livyatan melvillei*). Whether *L. melvillei* was white is not recorded in the fossil record, but it was certainly a great whale: it was twice the size of a modern killer whale and as large as any predator that has ever lived. (Melville would, one suspects, feel somewhat vindicated by all this, because the novel *Moby-Dick* was a commercial failure, and during his lifetime critics considered him only a minor figure in American letters.)

Rudyard Kipling's name lives on in a spider (*Bagheera kiplingi*), with a twist: its name honors both Kipling and one of his characters (the black panther Bagheera who befriends Mowgli in *The Jungle Book*). The film actors Kate Winslet and Arnold Schwarzenegger both have ground beetles (*Agra katewinsletae* and *Agra schwarzeneggeri*; Schwarzenegger's has suitably swollen biceps-like leg segments). Steven Spielberg has a pterosaur (*Coloborhynchus spielbergi*) and Pope John Paul II has a longhorn beetle (*Aegomorphus wojtylai*), and this may be the first time their names have occurred in the same sentence. The list goes on and on, with a new celebrity naming announced, it seems, every week.

Taxonomists (and other scientists) are divided about whether celebrity naming is a good idea. Some would like to reserve eponymous naming for scientists, arguing that whatever the merits of their music, Beyoncé and David Bowie (for example) have no connection to biology, so their names have no place in scientific nomenclature. Others take a stronger version of this position, holding that pop-culture celebrities simply don't deserve the adulation they receive, no matter how that adulation might be expressed. To these people, modern society is already too obsessed with people who aren't heroes but are simply doing their jobs—whether the job is hitting a baseball a long way, making funny jokes, or pretending to be a time-traveling cyborg assassin (Schwarzenegger in the movie *The Terminator,* for those who have forgotten). Those holding either of these views wouldn't suggest that scientists shouldn't be fans of Beyoncé or Ichiro—only that they should keep that fandom separate from their science. But why should they? Why must scientific names, or anything else in science, be depersonalized, serious, nothing more than functional? Why shouldn't scientists celebrate their passions—whatever those passions might be? Why can't *Scaptia beyonceae* show us that scientific names can be golden, sometimes, rather than gray?

A second argument against names like *Scaptia beyonceae* holds that celebrity eponyms trivialize naming, making the science of species discovery and biosystematics seem like unimportant play in the public eye. This is plausible, I suppose, although the counterargument is that celebrity naming is one of a very few things that brings species discovery into the public eye at all. Beyoncé's fly, for example, was written up in *Rolling Stone,* along with a beetle named for Roy Orbison, an isopod for Freddie Mercury, a roach for Jerry Garcia, a dinosaur for Mark Knopfler, and trilobites for Keith Richards, Paul Simon, Art Garfunkel, and all four Beatles. There's some overlap, of course, between the readerships of *Rolling Stone* and the *Australian Journal of Entomology;* but not that much.

Bringing horseflies, and species discovery, into the public eye is something that matters. In our modern world, governments slash funding for universities, museums, and basic scientific research, and taxonomy consistently gets the shortest end of all the short funding sticks. The museums that hold biological specimens are sometimes so underfunded that they can't keep their doors open, or protect their collections from disaster. In September 2018, the National Museum of Brazil was consumed by fire. Chronic underfunding played a major role in the loss—the museum, for example, lacked a functioning sprinkler system. Among the collections lost were five million insect specimens; among these specimens were, almost certainly, hundreds of new species waiting in their cabinets to be described and named. That's what had happened to *Scaptia beyonceae,* in its own museum, and there's nothing unusual at all about undescribed species in collections. Consider the case of Alexssandro Camargo's new robberfly, currently awaiting publication of its name. In 2018, Camargo recognized a specimen in London's Natural History Museum as representing a previously unknown species of the robberfly genus *Ichneumolaphria.* The specimen was collected in Brazil by Henry Walter Bates sometime during his 11-year collecting expedition there—an expedition that ended in 1859. The "new" *Ichneumolaphria* had spent 160 years, or perhaps a little more, in care of museum curators. It isn't alone: for Camargo's group of interest, robberflies of the New World tropics, the *average* "new" species has spent more than 50 years in a museum drawer somewhere. The typical voter, however, knows little about the collections departments of museums, which makes it very easy for governments to cut funding there instead of in more visible public services. The catastrophe of the Brazilian museum is, sadly, not unique: just two years earlier, the same thing had happened to the National Museum of Natural History in New Delhi, India. It would be foolish to think this can't, or won't, happen again. Against that backdrop, it's hard to criticize any effort to bring the science of taxonomy into the public eye.

A third criticism of celebrity naming is that it's an unseemly publicity stunt on the part of the namer—or even nothing more than an attempt to meet the eponymous celebrity. Could a taxonomist really think this might work? And might it? Frank Zappa (the experimental, and indescribable, musician) has at least five species named after him: a spider (*Pachygnatha zappa*), a mudskipper fish (*Zappa confluentus*), a fossil snail (*Amaurotoma zappa*), an enigmatic, so-far-unassignable fossil animal (*Spygoria zappania*), and a jellyfish (*Phialella zappai*). The namer of the last species, at least, did in fact have a cunning plan to meet Zappa. Ferdinando Boero, an Italian marine biologist, tells the story of heading to the Bodega Marine Laboratory in California to study jellyfish. Boero knew the eastern Pacific jellyfish fauna was poorly known, and he explained: "I would find some new species for sure. Once I had found them, I would have to give them . . . name[s]. I would dedicate one of them to Frank Zappa; I would tell him about it; he would invite me for a visit."[3]

Quite likely to Boero's surprise, the plan worked. He wrote to Zappa about the planned naming, and received a return letter from Zappa's wife, Gail, reporting Frank's reaction: "There's nothing I'd like better than having a jellyfish named after me."[4] The letter came with an invitation for Boero to visit Zappa at his home—which turned out to be the first of many visits over a long friendship. Zappa even dedicated his very last concert, in Genoa in 1988, to Boero, and sang a song about him. Does any of this make Boero's naming somehow disrespectful, or trivializing, of science? Was some damage done to taxonomy or invertebrate zoology? It's hard to see how. *P. zappai* needed a name, and it got one; Zappa has one more small legacy on Earth; and Boero has a story to tell. There's even a suggestion here, in Zappa's reaction to the idea, that people outside science can indeed be aware of the significance of a newly discovered species and the honor represented by its naming. That's an encouraging thing for the science of species discovery.

Some of the celebrities honored in species names are likely to have lasting fame, while others may ultimately prove to have been flashes in the pan. And that brings me to the final objection to scientists' indulgence in celebrity naming: that many such names are doomed to obscurity and etymological unhelpfulness. After all, with any luck, in a few years we'll all have forgotten who the Kardashians were. But surely the argument about the ephemeral nature of celebrity applies just as much to names in honor of anybody else. We already have thousands of species named for people who are now obscure (as becomes quite clear in the next chapter). At worst, these names become arbitrary: not one entomologist in a thousand knows which Smith the mosquito *Wyeomyia smithii* celebrates. At best, these names become maps to hidden treasure, rewarding those who follow the trail of clues with stories of fascinating people and human history. Perhaps in 100 years, someone will catch and identify Jon Stewart's wasp, and end up mystified by the bizarre twenty-first-century history that fed *The Daily Show.* Perhaps curiosity about *Heteropoda davidbowie* will lead someone to rediscover the music of David Bowie. Or perhaps a museum will mount a new fossil of the pterosaur *Coloborhynchus spielbergi,* and a search through some twenty-second-century descendent of Netflix will lead someone from *Jaws* to *Raiders of the Lost Ark* to *The Color Purple* and *Schindler's List.* That's a journey that will always be worth taking.

7

Spurlingia: A Snail for the Otherwise Forgotten

In the hot and scrubby forests of northern Queensland, in Australia, lives a small, brownish, obscure, and (to most people) uncharismatic land snail, *Spurlingia excellens.* This is one of a dozen Australian species assigned to a genus named *Spurlingia* in 1933. No other scientific name shares *Spurlingia*'s root. That would be extremely surprising if the name referred somehow to the snail's morphology, because those kinds of names tend to reoccur over and over again (there are probably thousands of species, for example, with names based on *rubra,* the Latin word for *red*). But "*Spurlingia*" has nothing to do with morphology. The snail's shell doesn't have spurs, and there's nothing particularly unusual about its tongue (Latin *lingua*). Instead, *Spurlingia* is another eponymous name—as you've no doubt guessed from its inclusion in this book. Who was Spurling, and what might the uniqueness of this name say about science and about the people who name species?

We owe the name *Spurlingia* to Tom Iredale (1880–1972). Iredale grew up as an avid birder and naturalist, but he had no university education. An invalid in his teens, he left his family and his native England at 21, traveling to New Zealand in search of a climate that would improve his health. Whether it was the climate there or just the change of scenery, he found himself restored to vigor. He worked as a business clerk, but spent his spare time rambling in the countryside and exploring New Zealand's natural history. Friends who joined his

Spurling's excellent land snail,
Spurlingia excellens

rambling got him interested in snails (peer pressure isn't always about misbehavior). After six years, life in business must have become unbearably stale, because in 1908 he joined an expedition to the Kermadec Islands—a sub-tropical archipelago 1,000 kilometers northeast of New Zealand. He spent ten months there, studying birds (and also shooting them to eat) and collecting snails. It was the end of his life in business, and the beginning of his life—despite the lack of a university degree—as a scientist.

Over the next two decades Iredale bounced back and forth across the world, eventually settling in Sydney, Australia, where he was hired as an assistant in the Shell Department at the Australian Museum. Before long, he was the museum's conchologist (head curator of snails and other molluscs), and he spent 20 years there collecting, studying, and writing about the molluscs and birds of Australasia. He was a prolific writer—so much so that he skipped dotting his i's and crossing his t's to save time. Over his career, he published more than 400 papers and named about 2,600 species and genera, including *Spurlingia*. It's no surprise, given this fertile career, that Irelade has his own collection of eponymous species—with dozens of molluscs and not a few birds being named *iredalei* and the like. Among the birds is an Australian warbler-like bird, the slender-billed thornbill (*Acanthiza iredalei*). *Acanthiza iredalei* is a good avian match for *Spurlingia*: a small brownish bird, uncharismatic except to the most devoted birders. Yet there's still honor in the name.

In naming *Spurlingia,* Iredale chose to honor an otherwise forgotten collector of shells and other natural history specimens named William Spurling. An obscure snail named for an obscure collector sounds like a recipe for a dull story, but in fact Spurling's story is anything but dull. It also has lessons to teach us.

William Spurling was indirectly affiliated with the great Victorian ornithologist John Gould, whose highly influential career spanned the middle decades of the 1800s. Gould remains well known today (at least among biologists) for two things. First, he produced a series of brilliant monographs on birds—among them, *The Birds of Australia, The Birds of Great Britain, The Birds of Europe,* and *The Birds of Papua New Guinea.* Second, it was Gould who received and identified Darwin's bird collections from the voyage of the *Beagle.* That includes the famous birds, collected in the Galápagos Islands, that we now call Darwin's finches. It was Gould, not Darwin, who realized that these birds were all closely related despite their conspicuous differences in bill shape, feeding behavior, and other traits. Darwin had thought them a hodgepodge of blackbirds, finches, and members of other bird groups, and didn't realize the significance of their similarities and differences until much later, after Gould's identifications. As a result, they aren't mentioned in *On the Origin of Species,* but they would later become a canonical example of the power of natural selection to shape organisms and to forge biodiversity. In a nice example of the unreliability of common names, by the way, Darwin's finches aren't actually finches. They're tanagers.

Besides Darwin's finches, Gould received specimens to examine from a network of collectors around the globe. Among these was Frederick Strange, who was born in England and collected in Australia. His life is, otherwise, rather sketchily reported. One brief biography includes a prominent assertion that he was semi-literate and financially incompetent, which can serve as a reminder that we do not get to choose the things for which we are remembered. He was certainly an accomplished collector and naturalist. Strange collected the first specimens of the Prince Albert lyrebird (*Menura alberti*)—a pheasant-sized bird whose secretive habits had let it go unsighted through 40 years of European exploration of eastern Australia. (It was named by the French ornithologist Charles-Lucien Bonaparte, and the bird had no connection to Prince Albert, consort of England's Queen Victoria, other than political expediency.) Among the birds and mammals, Strange's specimens also represented the scientific discoveries of the sooty owl, the plumed frogmouth, the mangrove honeyeater, the hoary wattled bat, the lesser stick-nest rat, and more. He was even more active as a collector of insects and especially of molluscs, although specimens of these less charismatic species were more likely to be traded, and used for species descriptions, without clear documentation of his role as collector. Nonetheless, quite a few snail species now bear the name *strangei*.

In August 1854, Strange bought a boat (the *Vision*, a 50-ton ketch), and the next month he sailed with a crew of nine for what was planned as a three- to four-month collecting expedition along the northeastern coast of Australia. His first stop was Curtis Island, near what is now Gladstone; his second was on Middle Percy Island (south of Mackay) on October 14th. The ship needed to replenish its water supplies, and at the same time, there were collections to be made. Strange took a party ashore on Middle Percy, including the expedition's botanist Walter Hill, an Indigenous Australian man named Deliapee, and three assistants: Henry Gittings, Richard Spinks, and William Spurling. Spurling's exact role in the collecting is unrecorded; he was listed as the ship's "mate," but shipboard job descriptions were often very flexible. Whether or not Spurling actually picked up specimens himself, he certainly joined shore parties to collect.

The visit to Middle Percy proved unfortunate. The party met a group of Indigenous Australians, and attempted to trade with them. They later complained that the Indigenous men failed to understand them (perhaps having had the gall to speak their own language rather than the Queen's English). Hill left the others to visit the highlands, and when he returned at the end of the day he found Spurling lying dead, with his throat cut, in some mangroves. Gittings, Spinks, and Strange were missing and presumed also killed (Deliapee later reported seeing Indigenous men attacking Strange with spears). Strange, the senior member of the party, was just 35 years old. Gittings was 20, and Spinks and Spurling were probably not much older.

Nobody now can have any idea *why* the four collectors were killed, but previous European visitors to the area had not behaved at all well. Mistreatment of Indigenous Australians by British explorers and settlers is, of course, well known. To add a more local context, seven years previously the Royal Navy survey ship HMS *Rattlesnake* had called at Middle Percy as part of its voyage to survey northeastern Australia and New Guinea. The *Rattlesnake* carried John MacGillivray as ship's botanist and Thomas Henry Huxley as assistant ship's surgeon and marine naturalist (Huxley would later be dubbed "Darwin's Bulldog" for his fierce support of Darwin's ideas about evolution by natural selection). They stopped at Middle Percy to repair some damage to the vessel, and crew and naturalists went ashore. The naturalists didn't encounter inhabitants on Middle Percy, but MacGillivray described wells, fireplaces, and other signs of recent use by Indigenous people. As for the crew: they didn't tread lightly on the land. Instead, they somehow managed to start a brushfire, which burned for several days and charred nearly the entire island.

With these events in memory, it's certainly understandable if the Indigenous men who met Strange and his party felt threatened.

Nevertheless, the colonial authorities sent a ship, the HMS *Torch,* the next year to find those responsible for the collectors' deaths, or failing that, to find some scapegoats. The crew of the *Torch* interrogated a number of Indigenous men and women on Middle Percy, piecing together a quite implausible account of the supposed murders. (Among other things, the *Torch*'s commander claimed that the Indigenous men admitted to consuming the body of one of their compatriots, after he was shot by Strange, and then hiding the bones on the island. This accusation says more about nineteenth-century European attitudes toward Indigenous people than it does about the truth of what happened.) In any event, three Indigenous men were arrested and sent to Sydney for trial, along with three women and four children. Rather surprisingly, the men were acquitted for lack of evidence—the surprise being not the lack of evidence, of course, but the acquittal. Sadly and *not* surprisingly, it seems likely that they died before being returned to their homes. As for the collectors: Spurling's body was recovered and buried at Port Curtis, Queensland, but the fate of the others is obscure.

History has preserved glimpses—but nothing more—of the members of Frederick Strange's ill-fated expedition. One of those glimpses is a rather terrible poem, "To the Memory of Frederick Strange, the Naturalist," published in 1874 by George French Angas, an Australian artist and naturalist. The poem opens with these stanzas:

> Australia! For thine onward march
> How many a son of science fell—
> Shall not thy bards of future years
> Their deeds of noble daring tell?
>
> Immortal they! Though wealth and pride
> Shall leave behind them scarce a name

The men who have for science died
Shall wear a changeless wreath of fame!

In fact, the "changeless wreath of fame" eluded Strange and, even more so, the expedition's other casualties. Of Spurling, Gittings, and Spinks, we have nothing now but a few newspaper mentions, a single tombstone (Spurling's), and *Spurlingia*.

What of Tom Iredale, who named *Spurlingia*? He was an impressive naturalist and conchologist, of course—naming 2,600 species makes that quite clear. But I think there was more to Iredale than that. In naming *Spurlingia,* he showed that he understood something about science and its progress that often eludes people. Our picture of the Victorian naturalists who scoured the globe for new species and new understanding tends to be a picture of Darwin, Bates, Wallace, Gould, and others: European or American men (nearly always men) who came from educated or at least prosperous backgrounds, had long careers, and left written accounts of their travels. That picture, though, is woefully inadequate. For every Darwin or Bates, Wallace or Gould, there were likely dozens of Spurlings: unheralded, likely uneducated, and largely unremembered. Some considered themselves scientists, others were hobbyists, and some played supporting roles—as guides, crew members, even cooks and craftsmen without whom the expeditions wouldn't have been possible. There were surely also dozens of Deliapees: local assistants who, as non-Europeans, weren't considered worthy of much comment (more on this later). It's true that Darwin and Wallace, for example, are remembered in part because they made outsized contributions to the development of science. But they didn't do that alone, and it seems fitting for us to remember the Spurlings too. Iredale's naming of *Spurlingia* does just that.

8

The Name of Evil

With very few exceptions, eponymous Latin names are intended to honor their eponyms. The naming is an act of respect on the part of the namer, and it's at least implicitly presumed that the eponymous person behind the name is worthy of that respect. There's no committee to review proposed eponymous names, though, to be sure this presumption holds. Sometimes, it doesn't.

In a few damp caves in Slovenia, a small, brownish, and rather dull beetle scrabbles in cave-floor sediment, hunting, in the dark, for the even smaller insects upon which it feeds. It's done that for thousands of years, while outside the caves human civilizations have risen and fallen, battles have raged, and empires have come and gone. The beetle knows nothing of human history, of course; and it doesn't know the name it's been given. If it did, though, it might not be pleased: its name is *Anophthalmus hitleri*.

Anophthalmus isn't the problem: that just means "eyeless." Like its 40 or so close relatives in the genus, and like many other cave animals that live in perpetual dark, *Anophthalmus hitleri* has neither eyes nor any need for them. But *hitleri* refers to *that* Hitler: Adolf Hitler, the führer of Nazi Germany, architect of the Holocaust, and as much the personification of evil as anyone who has ever lived. What could a tiny sightless beetle have done to deserve this—to deserve a name redolent of hatred, brutality, and mass murder on an unimaginable scale? Nothing, of course, except to have been discovered and de-scribed at the wrong time, by the wrong person.

Anophthalmus hitleri received its regrettable name in 1937, from an Austrian railway engineer and amateur entomologist named Oskar Scheibel. Scheibel received his first specimen of *A. hitleri* from a collector in Slovenia, and realized that it represented a species new to science. Scheibel was well positioned to make this realization. Despite his amateur status, he was an expert on cave beetles and on the group to which *A. hitleri* belongs (a subfamily of the ground beetles known as the Trechinae). His scientific judgment was just fine; his political judgment, considerably less so. He named *A. hitleri* in a short paper published in 1937, explaining that the name was "dedicated to Chancellor Adolf Hitler as an expression of my veneration" (in the original, "Dem herrn Reichskanzler Adolf Hitler als ausdruck mein verehrung zugeeignet").[1]

Little else seems to be known of Scheibel's life and views, and some have claimed that his admiration for Hitler was feigned for political purposes. One rather subtle clue, though, suggests that it was probably genuine. In the paper naming *A. hitleri,* Scheibel acknowledged the collector of the original specimen, one Lehrer Kodrič; but Kodrič's name appears with the parenthetical notation "(Gotschee)." Gottschee was an ethnically German enclave in southern Slovenia, so Scheibel was pointing out that Kodrič, despite his Slavic name, was of German blood. This could have no possible relevance to the collection or description of the beetle, which didn't come from the Gott-schee area. It's difficult, therefore, to imagine any reason for including that detail other than as a declaration of Scheibel's Aryanism—and that makes it difficult to see *A. hitleri*'s naming as anything other than more of the same. Even if that's wrong, though—even if Scheibel were only feigning that Aryanism—in the end it probably doesn't much matter. The name and Scheibel's professed reason for it were placed in the scientific record, and they remain there for all to see.

One would like to think that *Anophthalmus hitleri* was a one-time aberration, but unfortunately, it has company in its nomenclatural shame. A fossil insect (belonging to a long-extinct group, the Paleodictyoptera) was assigned the name *Rochlingia hitleri* in 1934—honoring not just Hitler but also the virulently anti-Semitic steel magnate Hermann Röchling. Hitler's Fascist Italian ally Benito Mussolini shows up too, in the blackberry *Rubus mussolinii.* Happily, *R. mussolinii* turns out to be just a variant of the common European blackberry *Rubus ulmifolia,* so the name *R. mussolinii* can be ignored as a junior synonym. No such luck for the two *hitleris.* Both remain accepted species, and under the principle of priority their names cannot be changed without a special ruling by the International Commission on Zoological Nomenclature. No formal petition for such a ruling seems to have been made, and it's unlikely that one would succeed. We're most likely stuck with *Anophthalmus* (and *Rochlingia*) *hitleri.*

It was, as we've seen, Linnaeus who introduced the possibility of eponymous naming, and therefore Linnaeus who made it possible for eponymous naming to go awry. To his credit, he anticipated the possibility, and was concerned about how it could be avoided. We can be less than impressed with his proposed solution. In his *Critica Botanica,* he argued that eponymous names "are to be bestowed in a proper manner, that is, chosen only by the greatest botanists. Hence this should be done by botanists of mature age, not by young men or newly hatched botanists: for in these kindles a sort of itch of enthusiasm, which is quenched in . . . riper age."[2] Younger botanists, you see, were too impulsive, too unreliable in judgment, to be trusted to confer eponymous names; in their enthusiasm, they might misstep. Of course, Linnaeus promptly, enthusiastically, and repeatedly indulged in eponymous naming himself, in the very same book and at the ripe old age of 30. Apparently, what Linnaeus meant by someone too young for the responsibility of eponymous naming was anyone younger than himself—an attitude that's betrayed almost every time someone proposes an age limit on anything. And what about Oskar Scheibel? When he named *Anophthalmus hitleri* in 1937, Scheibel was 56 years old.

Oddly, the temptation to name species (or genera) for despots and dictators seems to be difficult to resist. In addition to names for Hitler and Mussolini, there's a dinosaur genus *Jenghizkhan* (for Genghis Khan; although its members are now considered to belong to *Tyrannosaurus*). There's an African shrew *Crocidura attila* and a genus of silk moths *Caligula*. There's also an ichthyosaur *Leninia,* although the authors of the paper describing it insist that the name isn't actually in honor of Vladimir Lenin. Instead, the name is said to "reflect the geohistorical location of the find," because the type specimen is held in the Ulyanovsk Regional Museum of Local Lore, part of the Lenin Memorial and School complex in Ulyanovsk, Russia (Lenin's birthplace).[3] The proffered etymology is both unconvincing and easy to overlook, so most people will probably assume the name is eponymous. In effect, if most people assume that, it might as well be.

It's not only political leaders, of course, whom we might consider unsuitable for honoring in scientific names. Consider the Spanish conquistadors Hernán Cortés and Francisco Pizarro. Cortés led the Spanish expedition that overthrew the Aztec Empire in 1521, subjugating most of Mexico to the Spanish crown. Pizarro, a decade later, led the conquest of the Incan empire in Peru. Both men were brilliant tacticians, and for several centuries they were widely celebrated as heroes. To a more modern sensibility, these conquests are obviously regrettable exercises in colonialism, and their leaders committed war crimes. Yet both men are recognized in eponymous scientific names: Cortés in the beetle *Agathidium cortezi,* and Pizarro in the moth *Hellinsia pizarroi.* These aren't, as you might suppose, old names left to us from a more colonial day: *Agathidium cortezi* was named in 2005, and *H. pizarroi* in 2011. Interestingly, the more recent naming glosses over Pizarro's role as a colonial conqueror, referring to him as "the Spanish conquistador Francisco Pizarro, the first European to set foot in many parts of South America"—as if it was his presence, not his conquest, that mattered.[4] Those who named *A. cortezi* were a little more nuanced, referring to the "great Spanish explorer and conquistador Hernan Cortez who explored much of Mexico, conquered the local regime, and whose deeds and motivations remain somewhat controversial."[5] One might expect the naming to *also* remain "somewhat controversial"—as did the same scientists' naming of other *Agathidium* species, in the same monograph, for George W. Bush, Dick Cheney, and Donald Rumsfeld.

The potential for regret about naming is sometimes offered as a reason to avoid naming species for celebrities. After all, it's not unusual for an athlete, an actor, or a musician to be famous for professional accomplishment but to have questionable personal virtue. It's also not unusual for someone in that group to be widely admired for a long time, and then to make news because of misbehavior that had been previously hidden or ignored. The fact that no species have been named for Bill Cosby or Harvey Weinstein, to take two prominent examples, may reflect good luck as much as good judgment. But there are certainly names that raise questions. The trilobite *Arcticalymene viciousi,* for instance, is named for Sid Vicious, the Sex Pistols bassist and heroin addict who almost certainly murdered his partner Nancy Spungen. (Four other members of *Arcticalymene* are named for other members of the Sex Pistols: *A. cooki* for Paul Cook, *A. jonesi* for Steve Jones, *A. matlocki* for Glen Matlock, and *A. rotteni* for Johnny Rotten). The naming of the mite *Funkotriplogynium iagobadius* was undeniably clever (*iago* = James, *badius* = brown), but James Brown coupled brilliant musicianship and social activism with a long history of domestic and other violence. The act of naming a mite for James Brown might be seen as celebrating his music or his contributions to the civil rights movement; but it might also be seen as turning a blind eye to his persistent habit of assault.

The existence of unsavory personalities among the world's celebrities is no surprise. It would be terribly naive,

Cuvier's gazelle, *Gazella cuvieri*

though, to think that there haven't been unsavory characters among scientists, too. There have been, and we've named species after them. Consider, for example, the many species named for Georges Cuvier (1769–1832), Louis Agassiz (1807–1873), and Richard Owen (1804–1892): among them, Cuvier's gazelle, *Gazella cuvieri;* the Agassiz snowfly, *Isocapnia agassizi;* the little spotted kiwi, *Apteryx owenii.* Cuvier, Agassiz, and Owen each made enormous contributions to science, and the namings recognize those contributions. But each was unsavory, too. Owen was sloppy or worse in crediting the work of other scientists: in 1846 he was awarded a Royal Medal for a paper he'd written on belemnites (fossil squid-like animals), but he'd conveniently forgotten to mention Chaning Pearce, the naturalist who had discovered the first belemnite fossils four years earlier. Owen, whose vanity, arrogance, and vindictiveness had already worn his colleagues' patience thin, was voted off the Council of the Royal Society of London. As for Cuvier and Agassiz: each espoused racist views that might have been widespread in their own times but are revealed as shoddily reasoned and offensive today. Cuvier described three human races, arguing that "the white race, with oval face, long hair, and prominent nose, to which the civilized peoples of Europe belong, is the most beautiful and superior to the others by virtue of its intelligence, courage, and actions."[6] Agassiz, for his part, believed that the so-called races of humankind had been separately created by God, and upon meeting his first African American wrote to his mother of how he felt "pity for this degraded and degenerate race."[7]

This kind of thing isn't just a relic of centuries past. Until very recently, nobody had named a species for James Watson, one of the discoverers of the structure of DNA but also a man widely regarded as an unrepentant racist and sexist. That omission was arguably a fortunate one, but it didn't last. In March 2019, Watson got his species: the Indonesian weevil *Trigonopterus watsoni*. It's only a small consolation that *T. watsoni* is neither the largest nor the most handsome species of its genus. There are surely many more names like *T. watsoni* and *G. cuvieri;* after all, we have a very long list of eponymous scientific names commemorating thousands of botanists, zoologists, explorers, field collectors, lab technicians, and the like. It would be inconceivable if among them one couldn't find examples of most human failings.

What are we to make of the immortalization, via eponymous naming, of people whose conduct we might deplore? We might—we *should*—condemn James Brown's behavior with women and Georges Cuvier's beliefs about race. But does that mean we should refuse to speak their names, or more particularly that we should regret their recognition in the scientific names of species? The answer isn't obvious, but I'd suggest that it may not be such a bad thing to have some species named for people of questionable virtue. That's because the alternative is to divide the set of people for whom species might be named into a subset of unambiguously good people (worthy of eponymous immortalization) and another subset of unambiguously bad ones. That would be a naive way to think about people, because no such tidy subsetting is possible. Rather, humans exist all along the spectrum from the saintly to the unspeakably evil. Pretending otherwise opens to the door to a form of disingenuous hagiography in which we exaggerate someone's virtue to get them into the "good" column and then turn a blind eye to any imperfections in an attempt (conscious or unconscious) to keep them there. Not only that: our concept of how certain behaviors place people along the spectrum shifts across cultures and over time. How we deal with this is a complex matter of ongoing discussion, and matters far more broadly than scientific names: Rhodes scholarships, Wagner's operas, and Picasso's art, among many other things, all need this kind of thought. Eponymous names can honor, and they can serve as a reminder of dishonor, too; if they inspire the curious to learn more about the people the names recognize, perhaps they can be a reminder that most of us have a little bit of both saint and sinner.

Hitler, we'd all agree, stands apart. So does *Anophthalmus hitleri*. The world would, admittedly, be a better place if that particular cave beetle had been named something else. But how much better? It's true that enormous damage has been done to humanity by antisemitism, xenophobia, and all the other odious doctrines of the far right; unfortunately, there's no sign that we'll soon have left such idiocy behind us. But it's not obvious that the existence of the name *Anophthalmus hitleri* has made any real contribution to that damage. We should avoid such obviously dishonorable namings when we can, of course; but fear of a few namings gone wrong shouldn't drive us to abandon eponymy. There's too much opportunity for honor, for fascination, and for learning.

We can even learn something important from *Anophthalmus hitleri:* that scientists are human, just like everyone else, and therefore aren't immune from temptation or from evil. Oskar Scheibel, scientist (none the less for his amateur status) and human, decided that Hitler was worthy of honor. We can and we should repudiate that decision, but we should never forget that it could be made.

9

Richard Spruce and the Love of Liverworts

In the summer of 1854, a botanist named Richard Spruce lay in a hammock in Maipures, along the Orinoco River in eastern Colombia, gripped by the recurrent fevers of malaria. His fevers were eventually tamed (and his life saved) by doses of quinine: a drug extracted from the bark of trees in the South American genus *Cinchona*. Appropriately enough, before many more years passed Spruce would play a major role in bringing *Cinchona* trees from their native range in Ecuador into cultivation in India. This feat made quinine available cheaply around the world and saved millions of lives. It also raised the ire of South American governments, which worried (ultimately, quite correctly) that the European colonial powers would dominate quinine production, destroying their domestic industry. Whether the export of *Cinchona* represented humanitarianism, colonialism, or biopiracy—or perhaps, all three at once—Spruce's part in it was an amazing chapter in an even more amazing story. That's because Spruce's sickbed in Maipures was just one incident in an epic 15-year collecting trip around much of tropical South America. He covered thousands of kilometers of rivers and trails, endured every imaginable privation, and in addition to his *Cinchona* collections, he sent specimens of more than 7,000 plant species back to European botanists—hundreds of them proving to represent species new to science. The trip was a triumph; but it cost Spruce his health and very nearly (and more than once) his life.

Today, Richard Spruce is commemorated in the names of at least 200 plant species. Among them are *Podocarpus sprucei,* a timber tree; *Hevea spruceana,* a fungal-blight resistant rubber tree; *Picrolemma sprucei,* a shrub whose roots yield a potential new antimalarial drug; and *Passiflora sprucei, Oncidium sprucei, Aristolochia sprucei, Guzmania sprucei,* and *Bonellia sprucei,* tropical rain-forest vines, shrubs, and epiphytes with striking and showy flowers. These are beautiful plants and useful ones, and quite a few of them owe their recognition by Western science to collections Spruce made during his travels in the Amazonian rain forests. The names honor Spruce for his botanical contributions; and yet in a way they ring a bit hollow. You see, it seems Spruce didn't care all that much for showy plants or for useful ones. His real passion was for mosses and, especially, for liverworts. These are tiny plants, rarely showy, easily and frequently overlooked; but in Spruce, they inspired rapture. Fortunately, Spruce's name is also honored in the liverwort genera *Spruceanthus, Spruceina,* and *Sprucella,* and the mosses *Orthotrichum sprucei* and *Sorapilla sprucei* (among others).

Richard Spruce was born in Yorkshire in September 1817. He grew up rambling in the countryside, and as a teenager compiled detailed lists of the plant species he found near his home (reporting 403 species, for example, around his village of Ganthorpe). At the age of 22, he took a position as Mathematics Master at a school in York, but he hated the job—or at least, he hated everything about it except the long holidays that let him roam the countryside looking for plants. When the school closed a few years later, Spruce decided on a change of career—and quite a change it was. On the advice of some botanical colleagues, Spruce decided to become a professional plant collector. There was such a thing at the time, because wealthy collectors would pay to acquire specimens for their private herbaria; the opportunity

Spruce's bristle-moss, *Orthotrichum sprucei*

must have seemed perfect for Spruce. He set out for southwestern France, to collect in the Pyrenees, with his expenses paid by the botanist William Borrer in exchange for dibs on the first set of Spruce's collections. In the Pyrenees, Spruce collected broadly, but his passion for mosses and liverworts began to become obvious. The French naturalist Léon Dufour had reported 156 moss and 13 liverwort species from the area of Spruce's collections; Spruce extended that list to 386 mosses and 92 liverworts. (In a paper published shortly after his return from the Pyrenees, he did the same for the English valley of Teesdale, extending its moss flora from 4 species to 167.) Spruce's expedition to the Pyrenees lasted nearly a year, but it was just a quick jaunt compared with what would follow.

In 1849, Spruce built on the reputation he'd established in the Pyrenees to launch a similar, but hugely more ambitious, expedition to South America. Again the expedition would be funded by the sale of plant specimens to subscribers. This arrangement was set up in cooperation with two of England's most prominent botanists: William Hooker, the director of the Royal Botanic Gardens at Kew, and George Bentham, who as broker received Spruce's specimens and distributed them to the subscribers. Spruce arrived in July at Belém, in Brazil at the mouth of the Amazon, and spent three months there familiarizing himself with the climate and the ecology of the tropical forest. His first shipments of specimens back to Bentham were of such quality that the subscriber list doubled. It was an auspicious start, but of course Belém was a major city (capital of the Brazilian state of Pará), where it was easy to live and work. Spruce began his South American expedition with lodgings in the home of a wealthy merchant, with access to shops, fine cuisine, and lines of communication back to England through the busy shipping port. It must have felt quite comfortable—but that comfort wouldn't follow Spruce for long.

In October 1849, Spruce's adventures began in earnest. He traveled 750 kilometers up the Amazon to Santarém, a town of 2,000 people and then the largest settlement on the Amazon. From there Spruce pushed gradually deeper into the rain forests, making expeditions up tributaries of the Amazon and Rio Negro and into the Orinoco to reach some of the most inaccessible places in the Amazon basin. He covered thousands of kilometers along rivers and trails. It was never easy. For instance, in 1857 he decided to travel from Tarapoto to Baños, in eastern Ecuador. Today, this is a trip of 1,300 kilometers on a mediocre highway, taking perhaps 22 hours of driving. It took Spruce three months and left him "much fallen in flesh" and coughing blood. It was neither the first nor the last privation associated with Spruce's botanical collecting. It wasn't even the worst.

We expect stories of explorers to feature plenty of hardship and danger, and Spruce's travels through Amazonia and the Andes would meet these expectations in spades. Perhaps his first unpleasant adventure came a few days after Christmas in 1849, when he managed to get himself lost in the rain forest along the Trombetas River. First he left most of his guiding party; then he left behind his assistant, Robert

King; then he lost sight of his remaining guide and found himself alone in the forest. He eventually heard King calling for him, but couldn't find his guides—who, one imagines, may have been too busy rolling their eyes at the hopelessly naive and distractable Englishmen to hunt for them. It took the day and a good part of the night for Spruce and King to find their camp again, and "the effects of this disastrous journey hung on us for a full week. Besides the rheumatic pains and stiffness brought on by the wetting, our hands, feet, and legs were torn and thickly stuck with prickles, some of which produced ulcers. In comparison with these, the annoyance caused by the bites of ticks . . . and the stings of wasps and ants was trifling."[1] A few years later he would encounter ants with a sting rather more than "trifling," when he disturbed a nest of bullet ants. Stung repeatedly about his feet and ankles, he experienced pain he called "indescribable"—but then promptly described. It was, he wrote, pain as "that of a hundred thousand nettle-stings . . . my feet . . . trembled as though I had the palsy, and . . . perspiration ran down my face from the pain. With difficulty I repressed a strong inclination to vomit."[2] Amazonia's human inhabitants were dangerous, too. Not long before the bullet ants, Spruce watched as the inhabitants of San Carlos celebrated the Feast of St. John with a drunken riot. He spent the duration standing guard in the door of his house with a revolver in each hand. A year later, while working his way down the Rio Negro, he overheard his guides plotting to rob and kill him. This time, he spent the night awake in his canoe with a shotgun in his lap.

It was disease, though, that threatened Spruce the most. Just months into his expedition, Spruce counted himself fortunate to have left Belém before it experienced a major outbreak of yellow fever—but he nonetheless suffered attacks of constipation and "slow fever." In July of 1854, in Maipures on the Orinoco, he had his first brush with malaria: violent attacks of fever by night, unquenchable thirst, vomiting, difficulty breathing, and the inability to eat more than small amounts of arrowroot gruel. Spruce's guides expected his death, but after two weeks he had (at last) the presence of mind to start dosing himself with quinine. In all he suffered through 38 days of intense fevers before recovering enough to resume traveling; even three months later, he complained of weakness that interfered with his work. It wouldn't be the last time he came close to death.

Spruce didn't know it in 1854, but the quinine that saved his life would eventually become the focus of his collecting efforts, and the work for which he'd be best remembered. Quinine was extracted from the bark of trees in the genus *Cinchona,* and had been in use as a treatment for malaria for 200 years. The best *Cinchona* trees (with the highest quinine content) grew in remote parts of the Andes foothills, though, and were difficult to access and harvest. Worse, by the middle of the nineteenth century, global demand threatened to far outstrip supply, especially given mounting demand from the British military in Africa and India and from the Dutch colonies in east Asia. The most accessible groves of trees were being rapidly depleted, with little thought paid to reforestation. A century of attempts to collect and export seeds for cultivation elsewhere had come to nothing, but the European colonial powers were convinced that this was the only way they could cope with the scourge of malaria—and thus the only way they could maintain their empires. Yet such an acquisition was only becoming more difficult, as South American governments were beginning to restrict the export of living *Cinchona.* In England, in the late 1850s, the explorer Clements Markham secured funding from the India Office for expeditions to collect seeds and seedlings. Markham had traveled in South America and had even seen *Cinchona* groves, but he was no botanist. Richard Spruce was, and in what may have been the very best of Markham's quinine-related decisions, he settled on Spruce as a collector for the Ecuadorian *Cinchona* species.

When Spruce received the royal commission to collect *Cinchona*, late in 1859, he had been in South America for ten years. Given the hardships he'd endured over those years and the precarious state of his health, one could easily imagine that he'd be eager to get home to the Yorkshire dales. But he accepted the commission, by all evidence rather enthusiastically despite describing the job in a letter as "likely to occupy me (if my life be spared) for the best part of next year." He was well situated to take on the *Cinchona* collections: he'd been based for some time in the right region (at Ambato, Ecuador), he was familiar with the trees, and he had connections. In particular, he had befriended a man named James Taylor. Taylor was physician to General Juan José Flores, who had been the first president of Ecuador. Flores, in turn, controlled extensive tracts of *Cinchona*-rich forest, to which Spruce was able to negotiate access. Despite these advantages, Spruce's suspicion that the job would be difficult and dangerous was well founded.

Spruce spent the first half of 1860 prospecting the region around Ambato. He needed to be ready to collect seeds, and grow seedlings, when they were available in July. Although there were several species of *Cinchona* in the area, Spruce aimed to collect the red-barked *Cinchona succirubra* (now treated as *C. pubescens*), which was suspected to have the highest anti-malarial activity. A series of expeditions radiating from Ambato confirmed that the best *C. succirubra* forests remaining were on the western slopes of Chimborazo Mountain (finally bringing Spruce out of the Amazon basin and onto the *western* side of the Andes). Before Spruce reached Chimborazo, though, he suffered what he called in his journal "THE BREAKDOWN": he woke up one April morning paralyzed in his back and legs. He wrote later that "from that day forth, [he] was never able to sit straight up, or to walk about without great pain and discomfort." He was more or less bedridden for two months, but in June he did what he must have felt he had to do: drag himself upright and start out for the "Bark Forests of Chimborazo." That journey was only about 35 kilometers as the crow flies, but of course Spruce was no crow, and the trip brought him to "that state of prostration when to lie down quietly and die would have seemed a relief."[3] The trails crossed at least two passes of 12,000 feet elevation, with precipitous descents along trails that were narrow, muddy, and steep. Spruce recorded his pleasure at encountering only a little sleet, rather than snowstorms, on one of them; but he was less pleased with wind that "occasionally lifted small fragments of gravel and hurled them at us."[4] The trip took Spruce and his small party a week.

In mid-June, Spruce set up camp at a small settlement called Limón, from where he could access the red-barked *Cinchona*. Collecting seed still wasn't an easy task. For one thing, even here, the groves had been heavily exploited. Near the settlements, nearly every mature tree had been felled (although there were many sprouts from the stumps). The weather was cool and wet, too, so the seeds were ripening slowly. This was a problem because, expecting Spruce to buy the seeds, Limón's inhabitants were stripping the unripe fruits from the trees. Finally, Ecuador was in the midst of a civil war, and not long after Spruce's arrival troops of a faction based in Quito (and backed by Flores) began to march through Limón on their way to press attacks in the lowlands. They stripped the countryside of food (Spruce complained of losing access to the one patch of plantains he'd managed to find) and threatened to seize Spruce's horses, food, and other goods. Good news came in late July, though, in the form of Robert Cross: an expert gardener sent from Kew Gardens to help Spruce root *Cinchona* cuttings. They soon had thousands of young plants, but the days turned hot and the two men needed to spend hours carrying water in buckets to keep the plants alive. They were also the plants' only defense against marauding caterpillars.

By the second week of August, the *Cinchona* seeds finally began to ripen, and by early September Spruce had collected and dried 100,000 of them—from just ten trees at Limón and five more from another settlement some 20 kilometers away. But this was far from the end of the task: the cuttings and the seeds still needed to be transported to the coast for shipment to India. Fortunately, the civil war was ending, as forces allied with General Flores (from whose land Spruce was collecting *Cinchona*) took the city of Guayaquil. It was again safe (or at least safer) to travel, and at the end of September Spruce was able to make his way to Guayaquil. Cross remained at Limón with the growing cuttings until, at the end of November, he considered them well enough established to travel. The plants (637 of them) would be installed in "Wardian cases" (essentially, sealed wood and glass terraria) and rafted downriver from Limón to Guayaquil. In spite of heavy rains that swelled the rivers and made the three-day voyage "speedy but perilous," Spruce and his precious *Cinchona* made it to Guayaquil, and on January 2, 1861, the *Cinchona* seeds and cuttings left Ecuador on a steamer to Lima: the first leg of their journey to India. There, the seed would germinate, the cuttings would grow, and within 15 years there would be hundreds of thousands of *Cinchona succirubra* trees growing in Indian plantations.

His *Cinchona* commission complete, Spruce stayed three more years in South America, collecting as much as he could. He was greatly frustrated by his poor health: for most of that time, he could walk only a little, couldn't ride a horse, and was able to sit upright at a table only with great difficulty. As if this wasn't enough, he lost all his savings in a bank failure. In May 1864, realizing that he would never again be able to roam the rain forest, he returned to England. He spent the rest of his life in Yorkshire, living on a very modest government pension while doing as much botany as his health would allow. Perhaps surprisingly, he lived almost 30 more years, dying of influenza at the age of 76 in December 1893.

For a man frequently and persistently incapacitated, Spruce's scientific output was prodigious. Throughout his 15 years in South America, Spruce collected everything he could, and took notes on everything he couldn't. No plant escaped his notice, and he was conscious of (and conscientious about) the value of collecting and describing useful and ornamental plants. Spruce made copious notes on food plants, fiber plants, timber trees, medicinal plants, psychoactive plants, and several species of rubber tree (he was the first, in 1855, to publish a description of rubber harvesting and processing). He described beautiful plants too, many of which became highly sought after by collectors and gardeners, including orchids, passionflower, and the astonishing royal waterlily, with its floating leaves that can reach 3 meters in diameter. Spruce returned to England with thousands of pages of field and travel notes, detailing not only the botany of Amazonia and the Andes but also the region's geology, geography, and ethnography—but he never published them.

There was enough material in Spruce's notes for dozens of scientific papers, not to mention a travelogue that could have been a best seller. (Fifteen years after Spruce's death, in fact, a version of these notes compiled and edited by his friend Alfred Russel Wallace appeared as *Notes of a Botanist on the Amazon and Andes*. They make a fascinating read, and the first-person accounts of the expedition are vivid and gripping.) So why didn't Spruce publish them? Because he knew these detailed observations were important, but they weren't his passion. It was always the liverworts and mosses that sang to him, from his beginnings rambling in Yorkshire to the Pyrenees to his South American adventure. His publications after his return to England are dominated by works about these unassuming little plants. Most notable among them is his masterwork: the 600-page monograph *Hepaticae of the Amazon and the Andes of Peru and Ecuador* ("Hepaticae" is the technical name for liverworts). Nobody writes a 600-page monograph about anything without passion, of course, but we don't need to rely on this evidence to see how Spruce felt about his mosses and liverworts. We can, instead, read his own words. In a letter to his friend Daniel Hanbury, Spruce rhapsodized that "In equatorial plains, [Hepaticae] creep over the living leaves of bushes and ferns, and clothe them with a delicate tracery of silver-green, golden, or red-brown. . . . It is true that the Hepaticae have hardly as yet yielded any substance to man capable of stupefying him . . . nor are they good for food; but if man cannot torture them to his uses or abuses, they are infinitely more useful where God has placed them . . . and they are, at the least, useful to, and beautiful in, themselves—surely the primary motive for every individual existence."[5]

When Spruce's travels in South America brought him to the most abject misery, sick and exhausted and afraid, he turned again and again to the plants he loved the most: "Whenever rains, swollen streams, and grumbling Indians combined to overwhelm me with chagrin, I found reason to thank heaven which had enabled me to forget for the moment all my troubles in the contemplation of a simple moss."[6] And *Cinchona*, and rubber, and all those other economically useful plants Spruce painstakingly observed and described? He admitted the plants themselves could be handsome, but his letter to Hanbury made it clear: "When they are beaten to pulp or powder in the apothecary's mortar they lose most of their interest for me."[7]

So what of the plant species that carry Richard Spruce's name? Remember that these include a timber tree (*Podocarpus sprucei*), a rubber tree (*Hevea spruceana*), medicinal plants (including *Picro-lemma sprucei,* with its anti-malarial potential), and many spectacu-larly beautiful flowers (the passionflower *Passiflora sprucei,* the orchid *Oncidium sprucei,* the bromeliad *Guzmania sprucei,* and many more). In naming these species, scientists have honored Spruce for his out-sized contributions to tropical and economic botany. But these are the very plants that might be destined for the apothecary's mortar (liter-ally, in the case of *Picrolemma,* if perhaps metaphorically for the oth-ers). I'm sure Spruce wouldn't have resented any of these namings; he would have understood perfectly well that if you're a botanist study-ing *Passiflora,* then a naming in *Passiflora* is the honor you're able to bestow. But if *all* the species named for him were the useful ones and the beautiful ones and the familiar ones, one can imagine him feeling just a little let down, and perhaps just a little misunderstood.

Fortunately, that isn't the case. Among the mosses, there are (for example) *Leskea sprucei, Orthostichum sprucei,* and *Sorapilla sprucei;* and even more squarely in line with Spruce's passion, among the liv-erworts there are the genera *Spruceanthus, Spruceina,* and *Sprucella.* Dozens of mosses and liverworts bear Spruce's name, actually. Most of them were named after Spruce's death, including (oddly, and a bit sadly, all the liverworts); but a full fifteen moss species were named for him during his lifetime—beginning with *Leskea sprucei* and *Orthotri-chum sprucei* in 1845, when he was just leaving for his expedition to the Pyrenees, and ending with *Bryum sprucei* in 1875. Spruce knew, therefore, of the honor his colleagues were bestowing on him. He knew that sometimes, someone lost in the contemplation of a simple moss would be contemplating one named for him.

10

Names from the Ego

The fact that scientific names can be eponymous opens up an obviously tempting possibility for the discoverer of a new species: why not name it after yourself? If I were lucky enough to discover and describe a beautiful new bird-of-paradise, could I name it *Paradisaea heardii*? And if I could, should I?

I've been assured many times that such narcissistic self-tribute isn't allowed—but as it turns out, that isn't true. Neither the Botanical nor the Zoological Code disallows the practice, so my dreamed-of *Paradisaea heardii* would be a perfectly legitimate naming; science, and the bird, would be stuck with the name in perpetuity. A few scientists, in fact, *have* named species for themselves. Only a very few, though, and this seems to be because among taxonomists, self-naming is considered a major faux pas. It just isn't something one does; and when someone—occasionally—does do it, eyes roll.

The very earliest eye-rolling concerns none other than Carl Linnaeus, who (you'll remember) invented the binomial system that made self-naming possible. Linnaeus's favorite plant was a delicate forest-floor herb known in English as twinflower. (In Linnaeus's native Swedish, it's the rather less appealing *giktgräs,* or gout-grass.) Its scientific name? *Linnaea borealis,* and more technically, *Linnaea borealis* Linnaeus—the authority "Linnaeus" identifying the formal author of the name. It's no wonder that it's frequently said that Linnaeus

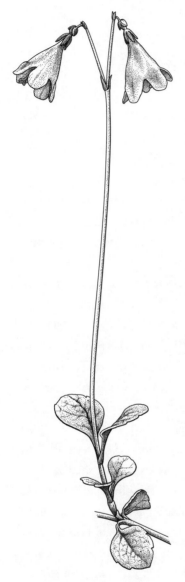

Twinflower, *Linnaea borealis*

named the genus *Linnaea* after himself. The real story isn't quite so simple.

Here's how *Linnaea* came to be called that. The name *Campanula serpyllifolia* had been in use for twinflower, and for a bunch of vaguely similar plants ranging from northernmost Europe south to the Mediterranean. Linnaeus, though, realized that the northern plants (twinflower) were not at all the same thing as the southern ones (although they have similar bell-shaped flowers, the northern and southern plants are now understood to belong to completely different plant families). The northern plants, therefore, needed a new name to distinguish them, and in his *Genera Plantarum* of 1737, Linnaeus gave them the name *Linnaea*. That sounds like self-naming—except that he noted that the name wasn't his own coining, but that of Jan Frederik Gronovius (an older Dutch botanist who was Linnaeus's patron and friend). So at least according to Linneaus, it was Gronovius who named twinflower *Linnaea*.

If Gronovius named *Linnaea,* why isn't the name's authority "Gronovius"? It would be, in any later era, but for such early namings there's a technicality that comes into play. The botanical Code specifies that the earliest possible authority for plant names is Linnaeus's *Species Plantarum* of 1753, and all names that appear there receive the authority "Linnaeus"—even if Linnaeus was only recording a name bestowed earlier by somebody else. And that's, at least on the face of it, what he was doing with Gronovius's name *Linnaea.*

So can we acquit Linnaeus of the charge of narcissistic self-naming? Digging deeper into the story suggests that we probably can't. In his 1971 biography of Linnaeus, Wilfrid Blunt suggested that Linnaeus arranged the whole thing, having "caused [the plant] to be renamed *Linnaea borealis* in his honor." Although Blunt made the accusation in passing, without evidence, it's most likely true. Linnaeus was by all accounts vain, with no tendency to undervalue his own contributions to botany; he surely felt himself deserving of the *Linnaea* tribute even if he had to orchestrate it. Consider this: a handwritten manuscript from 1730 for what would later (in 1736) become his *Fundamenta Botanica.* In that manuscript Linnaeus compares the morphological features of two plants: one labeled *Campanulam,* and the other *Linnaeam.* The comparison suggests that he had already decided that the northern sort of *Campanula serpyllifolia* (twinflower) was different from the southern sort and needed a new name, and that that new name should be *Linnaea* (or some variation on that). But given the timing, it's highly unlikely that the name really came from Gronovius. In 1730, Linnaeus was a 23-year-old medical student, in only his second year of study at the university in Uppsala. He was just beginning to build a reputation locally as a botanist, and was still five full years away from meeting Gronovius (when Linnaeus traveled to Holland). If, before 1730, Gronovius had heard of Linnaeus at all,

The smoking gun: "Linna'am" in Linnaeus's 1730 draft of *Fundamenta Botanica* (By permission of the Linnean Society of London)

he surely wouldn't have thought to name a plant after him—especially given that eponymous naming wasn't a common practice then. The case may well go back even another year. Another handwritten manuscript, this one of the *Spolia Plantarum* from 1729, gives northern *Campanula serpyllifolia* the name *Rudbeckia*. But there's an erasure just visible under *Rudbeckia*, and it looks a lot like *Linnaea*. It's entirely inconceivable that Gronovius would have named a plant for Linnaeus in or before 1729.

If Linnaeus was toying with the idea of *Linnaea* long before Gronovius could possibly have coined the name, then the later attribution of the name to Gronovius is almost certainly either disingenuous or a case of Linnaeus pulling the puppeteer's strings. We may never know for sure, but in light of the available historical clues, it's amusing to read his explanation of the name in yet another of his early works, the *Critica Botanica*. Here he tells us that *Linnaea* "was named by the celebrated Gronovius and is . . . lowly, insignificant, disregarded, flowering but for a brief space—from Linnaeus who resembles it."[1] The humility in this passage is palpably put-on. It's not hard to picture Linnaeus winking as he wrote it, tickled with his own cleverness in arranging, behind the scenes, for his favorite plant to bear his own name.

The story of Linnaeus arranging for *Linnaea's* naming has persisted since at least Blunt's original accusation, and it's done so (until now) without evidence. What's given the story this longevity? I suspect it's the fact that it's fun. There's something guiltily satisfying about the disapproval of someone else's conduct, and this psychological quirk shows up not just in modern-day social media but in the scientific literature too. That literature is sprinkled with accusations of self-naming—many of which have persisted despite turning out not to be true.

Consider, as an example, the tongue-twisting North African snail *Cecilioides bourguignatiana.* Its name honors one Jules-René Bourguignat, a French zoologist who named more than 2,500 species of mollusc (although his interests were much broader, and he wrote treatises on botany, geology, archaeology, and more). In 1864, he published a 500-page monograph on the molluscs of Algeria, in which he listed and described his eponymous snail (under the name *Ferussacia bourguignatiana*). An anonymous notice in the *American Journal of Conchology* praised Bourguignat's book as a "magnificent work," but added in a stiffly disapproving footnote: "We do not remember ever before to have seen an author name a species after himself, as Mr. Bourguignat has done in *Ferussacia bourguignatiana.*"[2] Bourguignat had many enemies and a reputation for arrogance, so it's not surprising that he was accused of naming *Ferussacia bourguignatiana* after himself. But the author of the disapproving footnote should perhaps have read a little more carefully, because Bourguignat wasn't responsible for the name at all. He was, instead, providing a new description (and a new genus name) for a snail christened *Achatina bourguignatiana*—two years earlier, by the Italian naturalist Luigi Benoit. No taxonomic infelicity was committed, but the story has persisted, unchallenged, for a century and a half (it's told, for example, in John Wright's *The Naming of the Shrew*).

Other supposed cases of self-naming turn out to involve confusion about pedigrees. Take, for example, the dung beetle *Cartwrightia cartwrighti*, named in 1967—suspiciously, to be sure—by entomologist Oscar Cartwright. Although the genus name *Cartwrightia* does indeed refer to Cartwright, it was named in his honor by *another* entomologist, Federico Islas Salas (who was honored in return by Cartwright, with *Cartwrightia islasi*). Cartwright was very pleased by the naming of *Cartwrightia* (and by beetles in 16 other genera that were named for him during his lifetime) but he wasn't responsible for it. And *Cartwrightia cartwrighti*? Cartwright did name that species, but he didn't name it for himself. Instead, *C. cartwrighti* honors his brother Raymond Cartwright, who was a frequent companion on the field trips that so often lie behind the discovery of new species. Improbable? Sabine's gull, *Larus sabini,* was named in 1818 by Joseph Sabine, in honour of *his* brother Edward Sabine. Edward had shot the first specimen, on one of many British expeditions in search of a Northwest Passage through the Canadian Arctic. Joseph, perhaps sensitive to the skepticism his naming might face, was careful to specify that the name *sabini* was "in conformity with the custom of affixing the name of the original discoverer to a species."[3]

Along similar lines, the nyala (a southern African antelope) *Tragelaphus angasii* was named by George French Angas in 1849, but not for himself; instead, the name *angasii* was "in honor of my esteemed father, George Fife Angas, Esq."[4] (Angas actually credited someone named "Gray," of the Zoological Society of London, with suggesting the name. It's not clear which of several zoological Grays this might have been, but in any event Angas was the first to publish the name, so he is the namer of record.) Confusion in cases like Cartwright's or Sabine's or Angas's is, I suppose, understandable. After all, there's an obvious conclusion one can jump to, and it's not always easy to decode the namer's real intentions. If the etymology of an eponymous name is explained at all, that explanation often lies buried deep in a technical paper—and often, a paper in an obscure publication that only the highly motivated (or the highly expert) will track down. Knowing this, a scientist wishing to name a species for a family member will often work with a given name rather than a surname— as, for example, with *Icius kumariae,* a jumping spider recently named by John Caleb for his (presumably not arachnophobic) wife Kumari.

Sometimes, self-naming is real but accidental. "Wait," you say, "how could someone possibly name a species after themselves *by accident?*" Here, again, the technicalities of naming come into play. The authority for a name comes from its earliest publication—and while that sounds simple enough, it introduces wrinkles that sometimes trip up an eager honoree. Take *Aphyosemion roloffi* Roloff. An amateur aquarist, Erhard Roloff, collected this killifish in western Africa in the 1930s, and recognized it as an undescribed species. He sent specimens to Ernst Ahl, an ichthyologist at Berlin's Natural History Museum, and Ahl prepared a description and the name *Aphyosemion roloffi*— honoring Roloff as the collector. There's nothing unusual about that, of course, and if Roloff had restrained himself the story would have ended there. He couldn't, though, and instead wrote an article for an aquarists' magazine mentioning the new species and its name. But circumstances delayed the publication of Ahl's paper until 1938, after Roloff's article appeared in 1936—making Roloff and not Ahl the namer by the rules of nomenclature. Roloff nearly did it again with *Rivulus roloffi,* another killifish (this time from Haiti). He sent his *Rivulus* specimens to Ethelwynn Trewavas at the British Museum, in 1938. The same story played out: he promptly published another magazine article mentioning the species and its name, while Trewavas's publication was delayed by the war until 1948. Perhaps fortunately, this time Roloff's description was too limited for his article to be recognized as the official naming. The name is sometimes given as *R. roloffi* Roloff—and there's that guilty pleasure in disapproval again—but it's really *R. roloffi* Trewavas, and Roloff, this time, is off the hook. It's easy enough to understand Roloff's eagerness to use the names of "his" new species; he was an avid hobbyist and must have been delighted by the honor of having not one but two killifish named for him. It's an amateur mistake, though, to jump the gun, and be the first (even accidentally) to use the name in print.

Like *Ferussacia bourguignatiana, Cartwrightia cartwrighti,* and *Rivulus roloffi,* most accusations of self-naming turn out to arise not from ego but from confusion. Where there's smoke, there isn't always fire. But sometimes there is. In 1785, Siegmund von Hochenwarth named a noctuid moth *Phalaena hochenwarthi* after himself; or at least, he left us to assume that, since he didn't explain his choice of name. Similarly, in 1937 the zoological collector and nature writer Ivan Sanderson mentioned, in passing and in a book for the popular market, a bat he referred to as *Hipposideros sandersoni.* Although he said nothing at all about the name's etymology, one can only assume he was staking a claim. The inference is clearer for *Rhinoceros jamrachi,* named by William Jamrach in 1875 based on a single living rhino shipped to England. Jamrach was a dealer in exotic animals, rather than a scientist, but he didn't hide his high opinion of his contributions to the field. He was sure that "his" rhino was a new and previously unknown species, and referring to his arguments with zoologists who disagreed, he wrote: "I stamped my foot and bestowed my blessings on science."[5] In case his readers remain skeptical, he ended his paper this way: "I rest contented with the idea of having named one of the THREE NEW SPECIES of Rhinoceroses brought to England, alive, by me."[6] So even though Jamrach doesn't explicitly *say* he's naming *R. jamrachi* for himself, it's pretty clear. Unfortunately for Jamrach, he was entirely wrong; "his" new rhinoceros was just another individual of the familiar Indian rhinoceros, named *Rhinoceros unicornis* by Linnaeus long before Jamrach's time. He got a second chance at nomenclatural immortality in 1906, with the cassowary *Casuarius jamrachi* (this time named for him by Walter Rothschild)—but that, too, was snatched away, as there proved to be no such species (the specimen most likely being a dwarf cassowary, *C. benneti*). The names *Rhinoceros jamrachi* and *Casuarius jamrachi* are now just discarded synonyms; and Jamrach, "his" rhinoceros, and "his" cassowary are now long forgotten. The only *jamrachi* name still in use is that of the snail *Amoria jamrachii*—named not for William Jamrach but for his father Charles. We can imagine William's infuriated foot-stamping.

Linnaeus's skulduggery and Jamrach's casual omission might give the impression that nobody ever comes right out and admits to naming a new species for him- or herself. That impression would be almost, but not quite, accurate. Cases of admitted self-naming are as rare as hens' teeth, but thanks to one Colonel Robert Tytler, there's at least one known example. Tytler was an officer in the Bengal Army in mid-1800s India, and a keen amateur naturalist who observed and collected birds, mammals, and reptiles. In 1864, he wrote a brief paper describing what he thought was a new species of civet from the Andaman Islands (like Jamrach's rhinoceros, Tytler's supposedly new civet didn't stand the test of time; it soon proved to be just the well-known masked palm civet, *Paguma larvata*). Wasting no time, he opened his paper this way:

As the mammalia found on these Islands must be of interest, I beg to send you the following description of a NEW *Paradoxurus* which I have named after myself,
PARADOXURUS TYTLERII[7]

Tytler, like many of the characters decorating this chapter, was an amateur. Perhaps he didn't know that among professional scientists self-naming was considered to be in poor taste. Or perhaps he felt the stereotyped self-assurance of the Imperial army officer, and simply didn't care what anyone else thought. Either way, in the name *Paradoxurus tytlerii* he left his intentions—and his ego—on display in black and white.

So: anyone *could* name a species after themselves, but it's a major faux pas, and with very rare exceptions, it just isn't done. This seems entirely appropriate, but also a little surprising. After all, scientists who name species are (despite the stereotypes) just as human as anybody else. There are shrinking violet scientists and there are blowhards; there are self-effacing scientists and there are blustering egotists; there are scientists who respect social norms and those who shred them; there are scientists who resist temptation, and those who succumb. It's curious that self-naming is a temptation that nearly (if not quite) everyone has resisted.

11

Eponymy Gone Wrong? Robert von Beringe's Gorilla and Dian Fossey's Tarsier

Eponymous Latin names often honor amazing people—preserving the memory of people we'd all agree are worth celebrating. But occasionally, it's only fair to recognize, they're duds. Consider, for example, two primates linked by the history encoded in their names. This is the story of a gorilla and a tarsier, of an old naming and a newer one, and of the possibility of nomenclatural regret.

In October 1902, Captain Friedrich Robert von Beringe climbed a volcano. Von Beringe was a German army officer and commander of an outpost at Bujumbura, in what was then German East Africa and is now Burundi. He had been traveling north to visit the king of Rwanda, and his party continued for an attempt to climb Mount Sabyinyo, one of the eight volcanoes making up the Virunga mountain chain. On October 17, they camped on a narrow, rocky ridge perhaps 500 meters below the summit. From that campsite, he sighted a group of what he called "big, black monkeys."[1] Von Beringe had become the first European to see a mountain gorilla. A few minutes later, he became the first European to kill one.

Von Beringe's party shot two gorillas, actually. Both bodies tumbled into a ravine, and it took several hours of labor to retrieve one of the two. Von Beringe recognized that he'd found an ape species unknown to science (although of course it was well known to the

The eastern mountain
gorilla, *Gorilla beringei
beringei*

Indigenous peoples of the area). It certainly wasn't a chimpanzee,
which he knew occurred in the lowlands nearby. It was also quite different from the western gorilla, and in any case Sabyinyo was more
than 1,000 kilometers distant from the range of that species. Von
Beringe sent the specimen to the Natural History Museum in
Berlin—minus the skin and one arm, lost along the way to a marauding hyaena. There it was studied by Paul Matschie, a zoologist, who
agreed with Von Beringe's assessment: this wasn't a chimpanzee or a
western gorilla; it was something new. He described it as a new species, and gave it the name *Gorilla beringei* in honor of its collector.

In October 1963, Dian Fossey climbed a volcano. Fossey was an
American occupational therapist who had worked for eight years at a
children's hospital in Louisville, Kentucky. She had always loved animals; occupational therapy was her second career choice, something
she settled on because she couldn't muster the grades to become a
veterinarian. After seeing photos of a friend's African safari, she re-
solved to travel there. It took her three years of planning and a loan

well in excess of her annual salary, but she was determined. She began in Kenya, engaging a guide who showed her elephants and rhinos and lions—exciting enough, but too mainstream to satisfy Fossey. Sources disagree about when and how she became obsessed with seeing the mountain gorilla, but she did; and when Fossey became obsessed with something, it generally happened. It took some serious browbeating of her guide, a difficult and dangerous trip into the Congo, and a grueling climb to the upper slopes of Mount Mikeno, but there she finally saw gorillas. She was utterly enchanted. Once home in Louisville, she became determined to return—and not just to see gorillas again, but to study them. Rather astonishingly, that's exactly what she did. With funding arranged by the famous anthropologist Louis Leakey, in January 1967 she set up camp back at Mount Mikeno, and began to observe the behavior and social structure of gorillas in the wild. Six months later, unrest in the Congo forced her to abandon Mount Mikeno (by some accounts, she narrowly escaped with her life), but she found a new site just a few kilometers away across the border into Rwanda. Here, at what she called the Karisoke Research Center, she spent most of the rest of her life.

Dian Fossey wasn't the first Western scientist to study mountain gorilla behavior (George Schaller, for example, had published both an academic monograph and a popular book about his work with the species). She wasn't particularly well qualified for the task, either. When she began her work, she was almost completely untrained in wildlife biology (she had audited a single course in primatology). She didn't speak any of the local languages or know much about the cultures, politics, or ecology of the area. In spite of all this, her influence was enormous, because she found ways to observe gorillas with an intimacy that others couldn't achieve. As Jane Goodall had done with the chimpanzees of Gombe, 200 kilometers south in Tanzania, Fossey let the gorillas see her and interact with her as they wished. She mimicked their activities, like feeding and scratching, and their vocalizations. Eventually, she gained their trust, and in doing so she was able to observe behavior that remained hidden to other researchers (at the risk, others would point out, of influencing the very behavior she was observing). She was also fanatically devoted to her gorillas. For 18 years, she spent as much time as she could at Karisoke; when she wasn't there, she was planning her return and working to maintain control over the work being done in her absence. Just as happened with Jane Goodall and her chimpanzees, Dian Fossey's name became inseparable from the gorillas she studied—in the public eye, in scientific circles, and among conservationists working to protect the gorillas and the rest of the African wildlife imperiled along with them. Fossey's work with her gorillas ended only when her life did, when she was brutally murdered in her cabin at Karisoke on December 27, 1985.

Dian Fossey's life and contributions to science have been marked in many ways. For a long time she was one of the most recognizable figures in science, partly as a result of repeated articles in *National Geographic*. Her book *Gorillas in the Mist* was a best seller, and her life and death were the subject of a major-studio movie by the same title, with Sigourney Weaver playing Fossey. Several biographies of Fossey have been written, beginning with Farley Mowat's *Virunga: The Passion of Dian Fossey*. None of these things, though, represent honors from Fossey's fellow scientists. Dian's tarsier, *Tarsius dianae*, does.

Tarsius dianae belongs to a group of small, nocturnal primates inhabiting the archipelagoes of southeast Asia. Tarsier taxonomy is difficult and controversial. Since the mid-1980s, the number of recognized species has grown from three to somewhere between a dozen and seventeen. Among these, *T. dianae* was named in 1991 by a group of German and French primatologists led by Carsten Niemitz. It occurs only on the island of Sulawesi, in Indonesia, where it lives in small groups in the rainforest feeding on insects and the occasional small vertebrate. Niemitz and colleagues gave two reasons for their choice of name: first, so that the "fierce little creatures" should bear the name of Diana, the Greek goddess of the hunt; and second, so that their species could honor Dian Fossey.[2] In a way, tarsiers are the perfect group of animals to have a species named for Fossey: tarsier species are morphologically rather similar, and distinctions between species rely to a great degree on differences in vocalizations. As a result, knowing tarsier species means following living animals in the wild, observing their behavior and listening to their calls—exactly the approach Fossey took to her gorillas.

Two primates, then, are connected by Dian Fossey's story: *Gorilla beringei,* the species she loved; and *Tarsius dianae,* the species given her name. Unfortunately, each demonstrates something of the drawbacks to eponymous naming.

Start with *Gorilla beringei.* Robert von Beringe, who shot that first mountain gorilla, was in Africa not as an explorer or a naturalist, but as part of the German military garrison in the colony of German East Africa. As a military man, he wasn't particularly successful. Perhaps surprisingly, given the genocidal history of European colonization in central and east Africa, he managed to be sacked by the colony's governor for being *too* violent (after a punitive raid he led against the Tutsi king of Burundi). There would be little reason to remember him today were it not for his accidental contribution to zoology. When he saw a "monkey" he didn't recognize high on the slopes of Mount Sabyinyo, his response was to shoot it. At least he had the presence of mind to recover the body and ship the specimen to the Berlin museum. That might seem to signal some interest in natural history, or awareness of the importance of science, but it's much more likely that von Beringe was just doing what colonialist Europeans did at the time: send unusual specimens back to their home countries, for the greater glory of empire. He seems to have made no other contribution to our knowledge of gorillas, or indeed to science more broadly. Nobody, one might argue, deserves eponymous immortalization less than Robert von Beringe.

What about *Tarsius dianae*? You might consider this name a dud for either of two reasons. First, the honor intended by it may prove to have been fleeting. Most primatologists now doubt that *T. dianae* is actually different from a tarsier described and named, from the same area in Sulawesi but in 1921, as *Tarsius dentatus*. If that's true, then the name *Tarsius dianae* is only a junior synonym and *T. dentatus* (because it's older) is the name to use. Unless future work reverses this judgment, seeing the honor in *T. dianae* will require a deep dive into the details of nomenclatural history in *Tarsius*. There's nothing unusual about this; Latin names are relegated to synonymy all the time. The loss of the eponymous name, though, seems to undercut the honor that was intended.

The second issue with the naming of *Tarsius dianae* is that many might ask whether the intended honor was really deserved. While Dian Fossey made important contributions to our knowledge of the mountain gorilla and to its conservation, by all accounts she wasn't a very pleasant person to be around. She was stubborn, abrasive, volatile, paranoid, sometimes violent, and seemingly racist. She assaulted suspected poachers and sometimes shot farmers' livestock. She bullied her staff, repeatedly sacking people or withholding their pay for perceived slights or deficiencies. She was often abusive toward students associated with the gorilla project. At one point, she wrote to a prospective student pointing out that (as of mid-1976) fifteen of eighteen students who joined the Karisoke project had abandoned it. She blamed their inability to cope with field conditions and suggested that they "had complexes of one kind or another"; it didn't seem to occur to her that her fits of screaming rage could be part of the problem.[3] In short, she made enemies easily and friends rarely; and her enemies would not be wrong to point to behavior that we'd agree today is unacceptable. Given all that, *should* her name be attached to a tarsier? This question (albeit not with specific reference to Fossey) is frequently raised as an objection to eponymous naming. We can take down a statue, or revoke an honorary degree; but we can't take back a Latin name. Perhaps for this reason we could cheer rather than rue the synonymy of *T. dianae*.

Fortunately, all is not lost. Names in both *Gorilla* and *Tarsier* have other stories to tell, and better ones. These can remind us that, just as the awfulness of the song "MacArthur Park" doesn't mean all music is a bad idea, a couple of duds don't compel us to regret Linnaeus's invention of eponymous naming. Let's take the two genera in turn.

Gorilla has only two species: the western *Gorilla gorilla* and the eastern *G. beringei*. However, *G. beringei* is divided into a pair of subspecies (geographically separated forms that aren't quite distinct enough to be considered different species). These subspecies have trinomial names: *Gorilla beringei beringei* for Fossey's eastern mountain gorilla, and *G. beringei graueri* for the eastern lowland gorilla. The latter subspecies name commemorates an Austrian mountaineer, explorer, and zoologist named Rudolf Grauer, who made repeated expeditions through northern and eastern Africa in the early 1900s. He sent thousands of specimens back to Vienna's Natural History Museum, including insects, birds, amphibians, reptiles, and a gorilla specimen that came to the attention of the same Paul Matschie who named *G. beringei*. That specimen became the basis for the name (Matschie described it as a species, *Gorilla graueri;* but then Matschie, who was resolutely opposed to the idea of evolution, thought every minor variant was a distinct species). Grauer has a number of other African species named for him, including a cuckoo-shrike (*Ceblepyris graueri*), a blind snake (*Letheobia graueri*), and a shrew (*Paracrocidura graueri*). These names honor Grauer for a contribution to our knowledge of African faunas far broader than von Beringe's—and one that resulted from an intentional approach to discovery rather than just an accidental encounter. Sadly, Grauer returned from his African expeditions with an unplanned collection: a variety of tropical diseases that plagued his health until he died in 1927.

Tarsius has more room than *Gorilla* for eponymy, with the recent recognition of a dozen or so new species. Three of them are eponymously named, and they celebrate 170 years of discovery of south Asian natural history. The first, *Tarsius wallacei,* honors Alfred Russel Wallace, one of the greatest naturalists of all time. Wallace spent eight years in the Malay Archipelago (from 1854 until 1862), and while there he collected more than 125,000 specimens—several thousand of which would prove to be species new to science. He made important contributions to the emerging science of biogeography (which is concerned with patterns in the geographic distribution of species), notably establishing the existence and importance of the geographic division between Asian and Australasian faunas, now called Wallace's Line in his honor. As if all that wasn't enough, during his Malay expedition he drafted a paper outlining his version of the theory of evolution by natural selection. That was a colossal accomplishment, overshadowed only by Charles Darwin's simultaneous co-discovery of the theory. *T. wallacei* is a good name, in the sense that Wallace richly deserves the honor. But because Wallace has plenty of other eponymously named species (more on this shortly), one might prefer two other tarsier names: *Tarsius spectrumgurskyae* and *T. supriatnai.* The first honors Sharon Gursky, a primatologist who has spent many years studying the behavior and vocalizations of tarsiers in the field. (The "spectrum" in *spectrumgurskyae* refers to an older name for the populations Gursky studied.) The second name honors Jatna Supriatna, an Indonesian herpetologist and primatologist who has made crucial contributions to tarsier conservation. Both Gursky and Supriatna's lives and careers have been profoundly intertwined with their namesake tarsiers, and both remain active in tarsier biology and conservation today. All of Asia's tarsiers face the threat of extinction, and scientists like Gursky and Supriatna are essential to their protection.

From von Beringe through Fossey to Gursky and Supriatna: the names of gorillas and the names of tarsiers tell us two things. First: our relatives among the apes have always fascinated us—for better, and sometimes for worse. And second: in our quest to understand their biology and ensure their survival in the wild, there's still plenty of work to be done.

12

Less Than a Tribute: The Temptation of Insult Naming

When Carl Linnaeus invented modern "binomial" Latin names, he made it possible for a scientist naming a new species to honor someone admirable or notable. But any tool that can build can also tear down; and just as Latin names can honor, they can dishonor. Linnaeus was the first to use naming to celebrate scientists who had gone before him—but he was also the first to succumb to temptation, and use Latin naming to insult someone with whom he had quarreled. He wouldn't be the last.

Linnaeus's most famous work, the *Systema Naturae,* used a new system for classifying plants: his "sexual system," in which plants were assigned to classes and orders based entirely on the number and arrangement of stamens and pistils in their flowers (stamens are the male, pollen-bearing structures, and pistils the female, ovule-bearing structures). For instance, his grouping "Octandria Monogynia" included plants with eight stamens and one pistil ("Octandria" from the Greek *octo* = eight, and *andros* = man, and "Monogynia" from the Greek *mono* = one, and *gyne* = woman). In places, Linneaus used somewhat bold language about this. For example, he described the Octandria Monogynia as "Eight men in the same bride's chamber, with one woman," and he explicitly equated the stigma to the vulva and the style to the vagina. He even waxed eloquently (and erotically, for the time) about how "petals do service as bridal beds . . . adorned with such noble bed curtains and perfumed with so many soft scents that the bridegroom with his bride might there celebrate their nuptials. . . . When now the bed is so prepared, it is time for the bridegroom to embrace his beloved bride and offer her his gifts."[1]

This sexual frankness didn't sit well with some of his contemporaries. A Prussian botanist named Johann Siegesbeck was particularly scandalized, condemning Linnaeus's system in a 1737 book as "lewd" and objecting to the notion that flowers could commit such "loathsome harlotry"—among other choice words. Siegesbeck and Linnaeus had previously enjoyed a friendly correspondence, but Linnaeus didn't take criticism well. He retaliated by naming a new species, *Sigesbeckia orientalis,* after Siegesbeck. How was this "retaliation"? *Sigesbeckia* is a small, unpleasantly sticky and rather unattractive weed, and one with tiny flowers to boot. Given Linnaeus's explicit association between plant and human sexual organs, his choice of a tiny-flowered species was surely no accident. In fact, the whole thing was far from subtle: earlier in the same year, Linnaeus had published his *Critica Botanica,* in which he set out explicitly the principles by which Latin names were to be constructed. Among these was that there should be a clear link, and preferably a resemblance, between the plant and the botanist it was named for. With this on record, it was impossible to miss the intended insult in *Sigesbeckia.* Or at least, impossible to miss it for long. Siegesbeck first thanked Linnaeus in a letter for having honored him with *Sigesbeckia*—but at that point, he was unfamiliar with the plant in question. Later, he came to understand, and Siegesbeck and Linnaeus would be enemies for the rest of their lives.

Linnaeus's sexual system didn't prove particularly useful as a way of classifying plants. Plant sexual anatomy *is* important, but just counting stamens and pistils doesn't get you very far. Before

Common St. Paul's wort, *Sigesbeckia orientalis*

long, the sexual system was largely abandoned—even by Linnaeus. It was replaced by various other systems that integrated information from many different traits in an attempt to organize plant diversity more naturally (and eventually, evolutionarily). Siegesbeck's objections, then, have become moot; but we still use the name *Sigesbeckia,* and *Sigesbeckia orientalis* is still weedy, sticky, and unattractive.

Linnaeus didn't actually come right out and say that he meant *Sigesbeckia* as an insult (although it was pretty hard to miss). He was a bit more up front (in *Critica Botanica,* his work from 1737) with a handful of other namings. Among them: *Pisonia,* a thorny and "sinister" tree for Willem Piso, whose work on Brazilian botany was sometimes held to be derivative of the earlier Georg Marcgraf; *Hernandia,* a tree with handsome leaves but inconspicuous flowers, for Francisco Hernández, whose work Linnnaeus judged ultimately unproductive; and *Dorstenia,* a group of mostly herbaceous relatives of mulberries, "whose flowers are not showy, as though they were faded and past their prime, [which] recalls the work of [Theodor] Dorsten."[2] Piso, Hernández, and Dorsten were all long dead when Linnaeus used naming to make his feelings clear, though. Only Siegesbeck was alive to feel the sting.

That Linnaeus let his insult of Siegesbeck require some reading between the lines isn't a big surprise, because he explained very few of his names. *Pisonia, Hernandia,* and *Dorstenia* are the exceptions, not the rule. Linnaeus wasn't unusual in this: at the time, almost nobody explained the names they bestowed. It didn't become routine until the twentieth century for taxonomists to explain the etymology of a new name (and even now etymology is only a recommended, not a required, part of a new species description). Even when etymologies are supplied, few taxonomists will explicitly record their intent to insult. One who did was Werner Greuter, and his target was a Czech botanist named Jiří Ponert. Ponert, in 1973, published a paper describing and naming 254 new plant species from Turkey. The botanical community was surprised by this, since Ponert was rather young and wasn't working in Turkey. It soon came to light that Ponert had simply taken descriptions of what seemed likely new species from a recently published Flora of the area, and assigned species names to them (copying the Flora's descriptions into Latin, standard practice for naming new species at the time). He had almost certainly never seen the specimens on which the descriptions were based. As it turns out, none of this contravenes the Code that governs plant naming, so Ponert's names are considered legitimate—but most scientists would consider his publishing practices dubious at best. Greuter commented on this quite creatively in 1976 by naming a Greek species of clover *Trifolium infamia-ponertii*—literally, the Clover of Ponert's Infamy. In a rather acerbic Latin footnote, Greuter explained that the name commemorates Ponert's inappropriate invention of names for plants he'd never seen.

Greuter's naming of *Trifolium infamia-ponertii* left his readers in no doubt about what he thought of Ponert. More often, understanding a namer's intent involves some careful reading between the lines and the compilation and integration of multiple clues. For example, two centuries after Linnaeus named *Sigesbeckia,* two paleontologists (Swedish again, although surely that's a coincidence) used their fossils to exchange particularly vicious naming insults. Understanding those insults involves some detective work.

Elsa Warburg and Orvar Isberg were invertebrate paleontologists working between the two world wars. Warburg was Jewish, and Isberg had far-right sympathies (during the Second World War he joined the pro-Nazi political party Svensk Opposition). Swedish paleontology was far too small a world for them not to interact frequently, and although little written record survives, it's clear that there was absolutely no love lost between them.

Warburg fired the first taxonomic shots. In her doctoral thesis in 1925, she named a genus of trilobites for Isberg. Although she graciously thanked Isberg for collecting the fossils she worked with, the naming was definitely not an honor. The new genus *Isbergia* contained two species, which Warburg named *Isbergia parvula* and *Isbergia planifrons.* Neither species name, in and of itself, is unusual; other trilobites bear similar names. However, in context, the meaning is only thinly veiled. The Latin *parvula* means slight, unimportant, or deficient in understanding, while *planifrons* means flat-headed. In light of Isberg's political beliefs, the latter name was particularly cutting (and Warburg made *I. planifrons* the type species for the genus). The far right believed that broad, flat head shapes were a sign of mental inferiority and associated them with "mediocre, inert" races (in the words of the French anthropologist Georges Vacher de Lapouge, whose work was eagerly seized upon by the Nazis). By associating Isberg with *planifrons,* Warburg was turning his own odious doctrine against him. The message couldn't be missed.

Nine years later, Isberg hit back, naming a genus of extinct mussels *Warburgia.* Just as Warburg had done in her own work, Isberg began with gracious thanks for his colleague's provision of specimens; but he left plenty of clues that these thanks were insincere. For one thing, Warburg was a woman of considerable size, and although Isberg named 20 new genera (several of them from specimens provided by Warburg), the genus he chose to bear her name had particularly "thick and full" shells.[3] In case that was too subtle, he described four species in *Warburgia: Warburgia crassa* (= fat), *Warburgia lata* (= wide), *Warburgia oviformis* (= egg-shaped), and *Warburgia iniqua* (= evil or unjust). These aren't very useful as descriptive names for the species, which don't differ much in shape; but the repetition in the first three names does its job. Finally, he points out that the genus is best distinguished from its close relatives by the obvious mark left by the anterior adductor muscle. What might that have to do with anything? Well, Isberg, writing in German, used the term *Schliessmuskel* for this muscle—a term that, in reference to humans, also means *sphincter* or *anus*. In the very next sentence he identifies Elsa Warburg as the eponym for the genus. By itself, each piece of the *Warburgia* description is unremarkable: some mussels are indeed fatter than others, many species bear names like *crassa* and *oviformis,* and there's no reason the Schliessmuskel can't be used to distinguish genera. Taking everything together, it's impossible to miss Isberg's intent.

I'm less convinced by *Dinohyus hollandi.* This extinct hog-like mammal (*Dinohyus* means "terrible hog") was named in 1905 by Olaf Peterson, a paleontologist at the Carnegie Museum in Pittsburgh, "in honor of Dr. W. J. Holland, the Director and Acting Curator of Paleontology in the Carnegie Museum."[4] A widely told story suggests that Peterson's naming was intended to slight Holland. Holland, the story goes, was known for insisting on being named the senior author of papers written by his staff, whether or not he had contributed to them (hogging the credit, hence "Holland, the terrible hog"). But we don't have strong evidence that Peterson intended the name as an insult, or that Holland took it as one. All accounts of the insult appear to be based on the same short passage in an unpublished autobiography by Robert Evan Sloan. Sloan, in turn, credits a story he was told by an older paleontologist, Bryan Patterson, who likely knew Peterson and Holland late in their careers but wasn't born until 4 years after *Dinohyus hollandi* was named. The *Dinohyus hollandi*-as-insult story is, then, at best third-hand. There are some reasons to doubt it, too. For one thing, it's not clear that Holland would have been insulted. He was a zoologist, paleontologist, and museum curator who had worked extensively with fossil mammals (even though his first love was butterflies and moths). Having *any* mammal species named after him seems likely to have pleased him. Furthermore, before publishing the name Peterson consulted with Holland, asking him to check the availability of the genus name *Dinohyus*. Holland suggested the alternative name *Dinochoerus*, but made no objection to *hollandi;* and in this exchange neither man's actions suggest an insult intended or understood. We may never be sure whether Peterson was trying to honor Holland or insult him, although it's fun to speculate.

Sometimes, a naming insult may be widely perceived even when it isn't intended. The best example of this may lie in the beetles *Agathidium bushi, Agathidium cheneyi,* and *Agathidium rumsfeldi.* They were named in 2005 by Kelly Miller and Quentin Wheeler, in a major revision that described 58 new species of this previously understudied genus. The names *bushi, cheneyi,* and *rumsfeldi* refer, of course, to George W. Bush, Dick Cheney, and Donald Rumsfeld—respectively president, vice president, and secretary of defense in the U.S. administration of the day. It was easy enough to think that these namings were meant as insults. After all, all three politicians were (and remain) reviled by many; the beetles feed on decaying fungi and the group is known as the "slime-mould beetles"; and another new naming, *Agathidium vaderi* (for the evil, but fictional, Sith Lord Darth Vader) was prominently illustrated as the monograph's frontispiece. But those seeing insults had jumped to a conclusion. The lead author, Kelly Miller, explains it this way: "We intended the names to be honorific. . . . We were two conservatives in academia working together (which is not common). It was early in the Iraq war period, and we were both in favor of intervention there. . . . And finally, we love our beetles! We wouldn't name a new species after someone we didn't like. [In interviews,] we compared it to the Lewis and Clark expedition naming the three forks of the Missouri after Jefferson, Madison, and Gallatin (President, Vice President, and Secretary of the Treasury [at the time])."[5]

Perhaps inevitably, there are those who believe nonetheless that *Agathidium bushi, cheneyi,* and *rumsfeldi* are cleverly disguised insults after all—pointing, as evidence, to *A. vaderi.* But in the same monograph, Miller and Wheeler named new species for their significant others past and present, for entomologists and collectors who had contributed to the study of *Agathidium,* and for their longtime scientific illustrator. It just doesn't make sense to believe that *Agathidium fawcetti* (for the illustrator) is an honor but *Agathidium bushi* is an insult.

Which brings us to Donald Trump. Few recent public figures have motivated such strong feelings among both supporters and detractors. During his run for the U.S. presidency, he was held up as everything from the savior of American freedom to its greatest threat. He also displayed a remarkable range of unpleasant (at least) behavior and proposed (and later instituted) policies, environmental and otherwise, that horrified most scientists. So the naming of a moth *Neopalpa donaldtrumpi,* in the first month of his administration, garnered considerable attention both in scientific circles and in the popular media. What might the namer, Vazrick Nazari, have meant to say?

On one level, it's quite obvious why the moth in question might be given the name *Neopalpa donaldtrumpi*. Its head is crowned with a remarkable shock of large, blond, comb-over scales, and Nazari points out its resemblance to Trump's equally remarkable hairstyle. But there is more to it than that. Nazari explained further this way: "The reason for this choice of name is to bring wider public attention to the need to continue protecting fragile habitats in the US that still contain many undescribed species."[6] *Neopalpa donaldtrumpi* is an appropriate species to call for that attention. It lives in dune habitats in California and Baja California (Mexico), and one of Trump's signature election proposals was to build a wall blocking (human) movement across the U.S.-Mexico border. The specimens on which Nazari based the species description were collected from the North Algodones Dunes Wilderness, a federally protected area. Another of Trump's promises was a push to roll back environmental regulation in the United States, and in particular to open up protected western land to exploitation. *Neopalpa donaldtrumpi,* therefore, can stand for the victims—both human and wild—of Trump's policies. Or, at least it can to those who disapprove of them. To some extent Nazari can be accused of preaching to the choir. After all, Trump's political base wasn't then, and may not ever be, likely to care much about a small brown moth or about its Latin name.

So is *Neopalpa donaldtrumpi* an insult naming? It was taken that way both by scientists and the media, for three main reasons. First, it's well known that scientists are disproportionately liberal in political outlook (remember Kelly Miller's comments in connection with *Agathidium bushi*), so one's first guess might always be opposition to Trump. Second, Trump's hair had been widely mocked, making any reference to it seem like it must be a jibe. Third—and to some eyes, most tellingly—Nazari's paper explains that *N. donaldtrumpi* is distinguished from its close relative *N. neonata* by the former's smaller male genitalia. This, of course, echoes common schoolboy taunts and seems to refer to Trump's assurance during the 2016 presidential debates that his hands (except, of course, he didn't mean his hands) were plenty big enough. But Nazari takes a more nuanced position, at least publicly. He insists that he based the name *donaldtrumpi* strictly on the moth's blond comb-over, and the size of its genitalia is merely a coincidence.[7] (He points out, quite correctly, that in insect taxonomy the size and shape of genitalia are routinely relied on as characters that distinguish species.) He stresses his desire to draw attention to the moth, its habitat, and its connection to Trump's policies, rather than to simply make fun of its eponym. He's not, actually, upset if people take the naming as an insult; neither is he upset if people take it as an honor. In fact, Nazari makes the interesting argument that the meaning of a name can be determined, in part, not only by its audience but by subsequent events. Should Trump's presidency, in hindsight, turn out to have been a positive force for the world, then he's content to have the name honor that. Should Trump's time in office leave the world diminished, the name can indeed be an insult. Time, Nazari suggests, will tell.

So: there have been a few clear cases of insult naming, and some other cases that are open to debate. Overall, though, the practice is uncommon. There are probably several reasons for this. First, at least in the Zoological Code, the recommendation that names "as far as possible . . . not cause offence" seems to discourage insult naming (although recommendations in the Code are not binding; and the Botanical Code, in contrast, explicitly permits "inappropriate or disagreeable" names). Second, whether or not insult naming is *allowed*, most taxonomists seem to agree that it's in poor taste. Finally, unless it's extremely pointed, it just doesn't seem likely to be that effective. If one's target is a scientist (or someone with strong interests in science), they're more likely to be pleased than insulted. And if they aren't, then scientific naming is probably too obscure for them to notice, or for the "insult" to sting if they do.

Except, perhaps, for *Isbergia planifrons,* and for *Sigesbeckia.* Those must have stung.

13

Charles Darwin's Tangled Bank

It is interesting to contemplate a tangled bank, clothed with many plants of many kinds, with birds singing on the bushes, with various insects flitting about, and with worms crawling through the damp earth, and to reflect that these elaborately constructed forms, so different from each other, and dependent on each other in so complex a manner, have all been produced by laws acting around us.

—*Charles Darwin,* On the Origin of Species, *1859*

Some eponymous names are unique: William Spurling has his snail genus *Spurlingia,* but no other organism on Earth carries his name. Others, like the 200 (at least) names of plant species and genera that honor Richard Spruce, have relatives aplenty. It's not a competition, of course—but if it were, who would win? Who has been most thoroughly commemorated in the names of Earth's flora and fauna?

The bad news first: this is a very difficult question to answer. Any attempt runs, rather rapidly, into two kinds of problem. The first: it's not simple to set the rules. If we want to count the species named for a particular person (as in my estimate of 200+ for Richard Spruce), what, exactly, should we include? Do we count every Spruce-eponymous name—every *sprucei, sprucella, spruceanum,* and the like—that's been published, or do we exclude those applied to species for which we now use a different name (that is, for which the *sprucei* name is considered a junior synonym)? Do we count the genus (of liverworts) *Sprucella* just once, or once for each species in the genus? Do we go beyond species to count names of orders, families, hybrid "species," subspecies, or varieties? What about cases when a *sprucei* name was published, but the publication didn't follow the rules of the Botanical Code and so is considered illegitimate? The weeds get thick quickly here.

But even if we could agree on all the rules, we'd have a second set of problems coming up with an accurate count. Efforts to compile all known Latin names into searchable databases have made great strides, but those databases can't yet be relied upon for exhaustive coverage— and the literature is far too vast to search from scratch. The databases are also full of ambiguities. You might, for example, wonder how many species are named for William Clark (co-leader of the Lewis and Clark expedition); but a database search for species with names containing the letter string "clark" will be of only limited help, because Clark's name is just too common. Tracking down the original literature sometimes helps, of course, but not always: until fairly recently it was disappointingly common for someone to publish a new species name without actually explaining its etymology. Sometimes there are clues—for example, if the species is described based on material collected on the Lewis and Clark expedition, *that* Clark is the most likely honoree—but often, it just isn't possible to discern the namer's intent.

Fortunately for our curiosity, there's good news along with the bad. Although it's a huge challenge to get exact numbers, a little work can yield estimates that should hold up well enough to settle the question. Who's been most thoroughly honored through eponymous Latin names? You can probably guess the answer—in fact, given this chapter's epigraph, you probably already have. It's almost certainly Charles Darwin, although the competition is tougher than you might have imagined.

Darwin's is surely the most instantly recognizable name in biology, and (with Isaac Newton, Albert Einstein, and a few others) one of the most recognizable names in all of science. His *On the Origin of Species* is the most famous scientific book ever published, and its concluding paragraph is the most famous passage in all of the scientific literature. That paragraph begins with the "tangled bank" account, which has persisted as a metaphor for Earth's biodiversity partly because of the fame of the book it closes, and partly because it's such a lovely piece of writing. The tangled bank is also a nice metaphor for the diversity of species bearing Darwin's name—species that include, fittingly, plants of many kinds, insects flitting about, and even worms crawling through the damp earth. Just how many species have been named for Darwin? A compilation of animal names by Miličić et al. (2011), together with some work trawling databases for plants, fungi, algae, and fossils, suggests an estimate of 363 species and 26 genera whose names honor Darwin. There are species named *darwini,* species named *darwiniana,* species named *charlesdarwini,* even species named *cephalidarwiniana* (literally and puzzlingly, "Darwin's head"). The naming began early, with a Chilean mouse, *Phyllotis darwini,* named in 1837 based on specimens collected by Darwin on the HMS *Beagle* expedition. It continues apace today: a recently named barnacle, for example, is *Regioscalpellum darwini,* and this species nicely acknowledges Darwin's much-less-famous work on the barnacles (he worked on the group for eight years, publishing four monographs).

Darwin's edge in the competition (which, remember, isn't a competition—but humor me, OK?) isn't huge. In addition to Richard Spruce, at least nine other people have over 200 eponymous

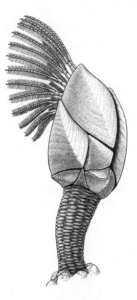

Darwin's gooseneck barnacle, *Regioscalpellum darwini*

names in their honor. Sadly, but unsurprisingly, they're all men and all of European ancestry, which is one small reflection of a problem science has not yet fully fixed. But in other ways, they're an interesting mix. Some of their names are nearly as familiar as Darwin's; others aren't, but should be. Some will never be well known outside their own fields. Some were born to great wealth and had the luxury of financing their own scientific careers, while others struggled to support themselves and their families. Who are these people who are nipping at Darwin's heels?

We can begin with Alfred Russel Wallace, Darwin's contemporary. A dozen years after Darwin returned from the *Beagle* expedition, Wallace left England for Brazil, where he spent ten years traveling and collecting in Amazonia (he met Richard Spruce there, and much later he would edit Spruce's journals for publication). He later spent eight years in the Malay Archipelago, where he arrived (independently) at the same insight that Darwin is famous for: that Earth's species have evolved from a single origin, with natural selection as evolution's primary mechanism. Although Darwin had the idea of natural selection first, and supported his argument more thoroughly and convincingly, Wallace deserves credit as its co-discoverer. Wallace also made huge contributions to biogeography, ecology, and environmental science, and more or less founded astrobiology (the science of life, if it exists, elsewhere in the universe) with his book *Man's Place in the Universe,* published in 1904. For 150 years, biologists have been naming species in Wallace's honor, in recognition of both his brilliance as a collector—he sent thousands of tropical specimens back to England—and his importance as a theorist. The result is at least 257 eponymously named species, and likely in the neighborhood of 300—short of Darwin's total, but nothing to sneeze at.

Another contender has equally close ties with Darwin: Joseph Dalton Hooker. Hooker, a botanist who became director of the Royal Botanic Garden at Kew, was a close friend and frequent correspondent of Darwin's. The two exchanged some 1,400 letters over a 40-year friendship. Early in his career, Hooker traveled the world collecting botanical and zoological specimens; his first voyage, in 1839, was to Antarctica on HMS *Erebus*. Other expeditions included the Himalayas, the Middle East, Morocco, and the United States (where he complained that "the beds are remarkably clean and good, but the pillows are too soft"). Later, he worked to maintain the Royal Botanic Garden's status as perhaps the most famous botanic garden in the world; he also built its herbarium into a stellar collection with specimens from every corner of the globe. Understanding of the world's flora grew by leaps and bounds at Kew under Hooker's direction. So how many species for Hooker? Ah, here the ambiguity problem rears its head, because Joseph Dalton Hooker's father (William Jackson Hooker) was *also* a brilliant and eminent botanist, *also* director of the Royal Botanic Garden (for the 24 years preceding his son), and *also* someone worthy of honoring with an eponymous species name. There are four or five hundred plants (and a few animals) with the name *hookeri* (or some variant on that)—but how many are for Hooker junior and how many for Hooker senior? Since many of the etymologies are unexplained, we'll probably never be sure; but the two Hookers were both titans of Victorian science, so the split can't be *too* lopsided. Let's credit them with 200–300 each.

We come next to Alexander von Humboldt. Humboldt died in May of 1859, just six months before the publication of *On the Origin of Species;* in that momentous year, one scientific era was beginning just as another came to a close. Humboldt was born in 1769 to a rich family, intimately connected to Prussia's aristocracy. As a boy he collected plants, insects, rocks—anything natural he could get his hands on—and at home he was called "the little apothecary" as a result. As a young man, he moved in Berlin's most learned scientific and philosophical circles, then attended a succession of universities to study economics, government, political science, mathematics, natural science, languages, finance, and (finally) geology and mining. He worked as a mines inspector, but was more interested in rambling the countryside to collect, conducting zoological experiments (he was fascinated with the effects of electricity on animals, both alive and dead), and discussing all manner of things with Prussia's intellectual elite: Johann Wolfgang von Goethe, Friedrich Schiller, and the like. But his dream was to travel the world. Milking his connections to the aristocracies of several European countries, he looked for a voyage to join, finally securing permission from the king of Spain for an expedition to Spanish territories in the New World. That voyage began in 1799 (exactly 100 years after Maria Sibylla Merian sailed for Suriname). Humboldt spent five years in the New World, there cementing his reputation as the most brilliant scientist of his time. Part of his genius was in using his observations to generalize about ecological patterns (like changes in vegetation along altitudinal and latitudinal gradients) that hold up worldwide. He was a polymath even by the standards of the time, over his long career publishing work in botany, zoology, mineralogy, ecology, geography, astronomy, political science, ethnography, and philosophy.

Humboldt's work influenced generations of scientists—including, notably in our current context, Darwin. As a young man yet to set foot on the *Beagle,* Darwin devoured Humboldt's accounts of his Latin American travels; he would later cite Humboldt as the inspiration for his own voyage. Humboldt isn't today the household name he was in his heyday. His name is nonetheless recognizable because it's been applied to geographical features all over, and beyond, our globe: mountains, for instance, on five continents; political units (town and counties) in ten U.S. states and one Canadian province; universities in at least two countries; and a sea—the Mare Humboldtianum—on the moon. And what about species (which are, after all, the reason for Humboldt's appearance in this chapter)? In *The Invention of Nature,* Andrea Wulf reports about 400 Humboldt names ("almost 300 plants and over 100 animals"), which would put Humboldt ahead of Darwin by a nose.[1] Wulf's count is almost certainly too liberal, though: getting to 400 seems to require counting names duplicated by taxonomic changes (for example, the names *Dumerilia humboldtii* and *Acourtia humboldtii* represent the same species of desert peony before and after a taxonomic revision moved it from genus to genus; one or the other should count, but not both). Correcting for this sort of thing gives a Humboldt estimate in the mid-200s—like Spruce's count and the Hookers', very respectable but well short of Darwin's 389. Perhaps Humboldt might have the most *things* named after him, if we count geography and the rest, but not the most *species.*

Three lesser-known botanists are up next: Augusto Weberbauer, Julian Steyermark, and the wonderfully named Cyrus Guernsey Pringle. We'll begin with Pringle (1838–1911), because he was in many ways Humboldt's diametrical opposite. Humboldt moved with Prussia's intellectual and social elite; Pringle grew up on a small farm in Vermont. Humboldt was born rich and was extensively educated; Pringle had his college career cut short because, after the deaths of his father and his older brother, he had to return to run the farm that supported him and his mother. Humboldt saw European wars around him, but was never touched by them; Pringle was drafted into the Union Army midway through the American Civil War (refusing to serve, as a pacifist Quaker, he was imprisoned and brutally mistreated before eventually being freed on the orders of Abraham Lincoln). Both Humboldt and Pringle loved plants, though. Pringle began to form his botanical interests breeding crop plants on his farm. In his mid-30s, he started to collect plant specimens around his native Vermont. He quickly built a reputation as a skilled collector, and in the 1880s was hired by Harvard University and the Smithsonian Institution to collect plants in the western United States and Mexico. Over the course of his travels he shipped some half-million specimens, representing 20,000 species, to herbaria around the world. Over 2,000 of those species would prove to be new to science, and about 300 of them now bear his name.

Weberbauer (1871–1948) and Steyermark (1909–1988) were similarly active plant collectors, and have about 250 eponymously named plant species each (and a small handful of animals between them). Weberbauer was born in Germany but spent most of his life living, teaching, and exploring for plants in Peru. In a pleasing connection, he taught for part of his career at the Colegio Peruano Alemán Alexander von Humboldt (the "Alexander von Humboldt German College of Peru"). Among Weberbauer's eponymous plants is a genus of spectacular columnar cacti from the Peruvian Andes, *Weberbauerocereus.* The cacti are spectacular, that is; the name is a little ponderous; and the doubly eponymous species *Weberbauerocereus weberbaueri* is worse. Finally, Steyermark (1909–1988) was an American who spent a long career collecting and studying the floras of Central and South America and of his home state of Missouri. His first work in South America found him following in the footsteps of Richard Spruce, hunting for *Cinchona* trees in Ecuador. In 1942, Japan had occupied

Java, the world's dominant source of quinine, and the United States military realized that malaria might be its worst enemy in the Pacific theater of the war. Steyermark joined more than two dozen other American botanists in the "*Cinchona* Missions" to locate native sources of *Cinchona* bark that could replace production from the Javan plantations. It was only the beginning of his South American collecting, which continued for 40 years and returned tens of thousands of specimens. Steyermark described and named about 2,000 species new to science, but that left plenty for other botanists to name in his honor.

Doesn't it seem odd that we haven't yet come to an entomologist? There are far more insect species than plant species needing names: perhaps half a million living plant species (400,000 of them named so far) compared to—well, we don't know quite what to compare it to. There are at least two million living insect species, quite likely ten million, and possibly as many as 100 million. Of course, only a little under a million of them have been described and named so far; but still, entomologists have had a lot to work with. Sure enough, at least two entomologists have joined the 200-eponyms club: Willy Kuschel and Geoffrey Monteith.

Willy Kuschel (1918–2017) worked in Chile and New Zealand studying the weevils (which are rivaled only by the rove beetles for diversity among insect families). Kuschel grew up on his family's farm in southern Chile, and came to the study of entomology in rather roundabout fashion. As a university student, he studied philosophy for two years, then theology for four, and spent two years as a priest before returning to school for a teaching degree. Entomology was his fourth discipline, and as an entomologist he earned the first Ph.D. degree awarded by the University of Chile, in 1953. He was an energetic and determined field collector, and built enormous collections of insects from the most inaccessible locations in South America and later in New Zealand and New Caledonia. Many of the species he collected now bear his name: 212 of them, by a recent count, along with 28 genera.

Geoffrey Monteith is the most recent name on my list of contenders, and the only one still active in science—and this makes his place in the field a bit surprising. Humboldt, Wallace, Pringle, and the rest have had a long time to accumulate eponymously named species. With the possible exception of Steyermark and Kuschel, they also worked at times when much of the world was just being opened up to Western biological exploration—exploration in which they all played notable parts. In comparison, Monteith is a babe in arms. He's an Australian entomologist, born in 1942, but already with 225 species and 15 genera named for him. This avalanche of namings seems, largely, to reflect two facets of Monteith's career. First, he's been curator of two of Australia's largest museum collections of insects and invertebrates, and in that position enthusiastically sent collections off to expert taxonomists who would sort and identify them—invariably discovering in those drawers and boxes species new to science, and often naming some of them for Monteith. Second, like Kuschel he collected thousands upon thousands of specimens himself, leading expeditions into the mountains of North Queensland and New Caledonia at a time when their faunas were virtually unknown to Western science—remnants of the scientifically untrampled ground the whole world had been when Wallace and Darwin and the Hookers were busily amassing their collections. As Monteith puts it:

I was a field-oriented biologist at a time when there were many unknown mountains to climb. I had a bunch of people over the years who absolutely loved, like me . . . busting our guts to get to new places, loved camping light to make room for collecting gear in our packs, loved squatting around a little fire under a nylon fly cooking our dinner while the rain sprayed in and soaked our bums . . . loved spraying mossy tree trunks and seeing an unknown fauna of tiny critters tumble down. . . . Every one of those very old tropical mountains in north Queensland had a whole unknown fauna of strange insects and arachnids. . . . And when we had almost exhausted those mountains the opportunity came to go to New Caledonia . . . and we found a similarly uncollected bunch of even higher, wet, tropical mountains stretching the 500 mile length of that bizarre and isolated island.[2]

Naming a new species for its collector is a common thing, and Monteith had the drive and the opportunity to collect a *lot* of new species.

And that's our pack of contenders—Darwin, and the ten close behind him. Perhaps you've notice a surprising omission. What about Linnaeus? He, after all, not only invented our system of scientific naming but, in doing so, made eponymous naming possible. His role in the science of taxonomy was foundational, to say the least; and on the other end of the transaction, he *gave* names to many thousands of species. And yet there seem to be "only" a hundred or so species named for him—not insignificant, to be sure, but trailing far behind much less famous names like Pringle and Monteith. It isn't obvious why Linnaeus hasn't been more widely honored. It may be because he wasn't particularly notable as a collector, working instead with specimens sent by a network of students and colleagues who traveled the globe while Linnaeus (mostly) remained in Uppsala. Or it may be because basing a name on Linnaeus is just a little too obvious.

Whatever the reason, Linnaeus (who couldn't imagine an honor he didn't deserve) would surely be nettled not to be in this race.

Where, exactly, does our survey leave us? If we simply count eponymously named species, there are at least a few close rivals to Darwin. But seen from a slightly different perspective, two names stand apart from all the rest: Darwin, of course, and Wallace. Richard Spruce has over 200 eponymous species, but they're all plants: from mosses to towering trees, to be sure, but nonetheless all plants. For that matter, all of Spruce's plants are extant (modern) species, a selection from the present-day veneer on the deeper history of life on Earth. The same is true, or nearly so, for Pringle, Weberbauer, Steyermark, and the Hookers junior and senior. On the other side of the tree of life, Willy Kuschel and Geoffrey Monteith's species are nearly all insects, spiders, and other arthropods. Only Wallace and Darwin have substantial numbers of living species and extinct ones, and species drawn from every branch of the tree of life on Earth.

Consider the 389 names honoring Darwin. They include plants (like Darwin's cotton, *Gossypium darwinii*), insects (Darwin's leafcutter bee, *Megachile darwiniana*), and worms (Darwin's earthworm, *Kynotus darwini*)—and there's your tangled bank. There are also algae (Darwin's coralline alga, *Lithothamnion darwini*), fungi (Darwin's webcap, *Cortinarius darwinii*), lichens (Darwin's reindeer lichen, *Cladonia darwinii*), sponges (Darwin's sponge, *Mycale darwini*), corals (Darwin's soft coral, *Pacifigorgia darwinii*), fish (Darwin's snailfish, *Paraliparis darwini*), frogs (Darwin's tree-hole frog, *Ingerana charlesdarwini*), lizards (Darwin's wall gecko, *Tarentola darwini*), mammals (Darwin's leaf-eared mouse, *Phyllotis darwinii*), and birds (Darwin's tinamou, *Nothura darwinii*). There's even a dinosaur (*Darwinsaurus evolutionis,* although that name's more clever than it is mellifluous). The list for Wallace is similar, albeit shorter. The species named for

Darwin and Wallace span both the breadth and the depth of Earth's biodiversity—and that's just right. After all, those two naturalists' stunning accomplishment was to unite our thinking about *all* life on Earth. Physics is still waiting for its Grand Unified Theory, but biology has had its for 160 years: the theory of evolution by natural selection. (Don't let the term "theory" here fool you; evolution by natural selection is as real as anything in our natural world, and considerably better understood than, say, gravity.)

There were evolutionary ideas before Darwin and Wallace; Humboldt, among others, had written about gradual transformations in the traits of species. But it was Darwin's work (complemented by Wallace's parallel ideas) that established evolution by natural selection as the underpinning framework for all of biology. It's evolution by natural selection that explains the remarkable differences between birds and bees, between fish and orchids, and between sponges and seaweeds. It's also evolution by natural selection that explains the remarkable similarities between them all. That's true for features Darwin and Wallace knew about, such as the repeated and convergent evolution of wings in flying animals. But even more impressively, it's true for features that were entirely unsuspected in Darwin's day, like the DNA-to-protein coding that all living organisms use (with only minor variation) to record and transmit genetic information. The fact that all living things share a common origin, and have diversified ever since through the process of natural selection, is what lets us understand all life on Earth as variations on a theme—astonishingly rich variation, to be sure, but with a theme that makes biology a synthetic science and not just a collection of special cases.

So while each of Darwin's eponymous species, by itself, represents a celebration of his importance to science (and ditto for Wallace), the greatest honor is in the whole tangled bank. Perhaps it's the wisdom of crowds, with taxonomists around the world, working across the tree of life and deep into its history, combining to fix Darwin's name to species that, together, suggest the diversity of life itself. That diversity is stunning, and the insight that evolution by natural selection underlies it all is no less so. Darwin knew just how important his insight was, and he summed it up in the famous conclusion to his "tangled bank" paragraph (and thus to *On the Origin of Species*): "There is grandeur in this view of life, with its several powers, having been originally breathed into a few forms or into one; and that, whilst this planet has gone cycling on according to the fixed law of gravity, from so simple a beginning endless forms most beautiful and most wonderful have been, and are being, evolved."[3]

Endless forms most beautiful and wonderful, and all of them needing names. Darwin's share *should* be the tangled bank it is.

14

Love in a Latin Name

"How do I love thee?" asked Elizabeth Barrett Browning, and answered, "Let me count the ways." That line, so familiar as to be perilously close to cliché, opens the 43rd of her *Sonnets from the Portuguese.* Barrett Browning wrote those sonnets for and about her husband, Robert Browning, turning to her professional tools as a poet to express her love. Turning to *their* professional tools, Pablo Picasso painted at least 60 portraits of his (first) lover, Fernande Olivier, and Richard Wagner wrote his *Siegfried Idyll* for his (second) wife Cosima.

That love can be recorded in poetry, or painting, or music will surprise no one. We're accustomed to the arts being used to explore and to record emotion. But what about science? It's a popular belief that there's no place for emotion in science, and that the scientist should therefore be coldly clinical and value detachment and objectivity above all else. It may well be true that emotion shouldn't drive the conclusions a scientist draws; but it's certainly *not* true that scientists are emotionless in their work. They can show that emotion in choosing what research questions to ask, in writing or talking about what they've done—or in the case of scientists who discover new species, in giving those species names. And if love is the greatest of all human emotions, it's reassuring that poets and painters and musicians don't have a monopoly on its expression. Scientists have named new species for their daughters and their sons, their sisters and their brothers, their wives and their husbands, even—sometimes—their unrequited crushes and their clandestine lovers. If insult naming shows scientists succumbing to their worst impulses, perhaps recording love in Latin names shows scientists as humans at their best.

Names in honor of children are quite common. We can begin with Charles-Lucien Bonaparte, a nephew of Napoleon Bonaparte. He was a French aristocrat who became an Italian prince, but he was also a biologist and ornithologist who discovered and named a number of new bird species. (He also "discovered" and promoted a young American naturalist named John James Audubon, whose paintings of American birds later became famous—but whose career was held back because he lacked connections with the aristocracy of American science.) In 1854, Bonaparte described a new imperial pigeon from the Philippines, naming it *Ptilocolpa carola* (it's now *Ducula carola*). The species name, *carola,* is a Latinization of Charlotte, for Bonaparte's 22-year-old daughter. Bonaparte wrote, "I dedicate it [the name] to my daughter the countess Primoli, Charlotte, herself worthy of her illustrious name," which is touching but also a bit odd.[1] It's odd because Bonaparte refers to his daughter as "the countess" and trumpets her family connections (she shared the name Charlotte with her aunt, a princess and niece of Napoleon)—yet Bonaparte was a staunch republican who decried the tendency for others to name species after royalty. In fact, four years earlier, he had named a bird of paradise *Diphyllodes respublica* as a statement of his republican ideals. Love, famously, is blind; and in this case, Bonaparte's love for his daughter may have blinded him a little to the irony of "Charlotte, the countess Primoli."

Birds for daughters seems to have been something of a trend. In 1846, Jules Bourcier (with his co-author Etienne Mulsant) named a hummingbird *Trochilus franciae* for his daughter Francia. In 1902,

Francia's hummingbird, *Amazilia franciae*

Otto Finsch named a southeast Asian finch (a coincidence, as the German word for *finch* is *fink*, not *finsch*) *Serinus estherae* after his daughter Esther. In perhaps the most poignant example, in 1839 René Lesson named a new species of mynah bird *Sericulus anais* in memory of his daughter, writing, "[for] Anais Lesson, deceased at the age of 11 years; may the name of this bird remember a father's deepest sorrow."[2] All three birds were later reassigned to different genera—they are now *Amazilia franciae*, *Chrysocorythus estherae*, and *Mino anais*—but the daughters' names remain.

It's not just birds, though, and not just the nineteenth century. A small Australian dinosaur, *Leaellynasaura amicagraphica*, was named in 1989 by Thomas and Patricia Rich for their daughter, Leaellyn. If there's anything more exciting for a child than having a dinosaur named after you, it can only be helping to dig it up in the first place; and sure enough, Leaellyn, as a schoolgirl, was part of the fossil's discovery. How could any child not be envious? A few years later, Leaellyn's brother Timothy got a dinosaur too, in *Timimus*, named jointly for the Riches' son and the Australian scientist and environmentalist Tim Flannery. But of course not every parent studies dinosaurs (or birds), and some children thus receive stranger tributes. Judith Winston's daughter Eliza got a bryozoan—a "moss animal." Bryozoans are tiny, colonial aquatic invertebrates that bear a superficial resemblance to corals. The new species *Noella elizae* was right for Eliza, Winston explained, because its orange tentacles matched Eliza's strawberry-blonde hair. "I don't know if she'll forgive me or not," wondered Winston in a media interview.[3] *Noella elizae* got its name only in 2014, so it's probably too soon to know the answer to that question. Children often find parental love embarrassing for a while, but grow into appreciation later.

Spouses make frequent appearances in the names of species too. Charles-Lucien Bonaparte named a genus of doves after his wife Zénaïde (mother, of course, of Charlotte, the countess Primoli): it's *Zenaida,* to which the familiar North American mourning and white-winged doves belong. Not to be outdone, Jules Bourcier named a hummingbird for his wife Aline (*Ornismya alinae*), and René Lesson named a hummingbird and an imperial pigeon for his two wives Clémence and Zoë (*Lampornis clemenciae* and *Columba zoeae*). It's probably not surprising to see nineteenth-century names honoring wives rather than husbands, because until very recently, it was almost always men doing the naming. One large set of species for which eponymous names have been exhaustively tabulated is the plant genus *Aloe,* and that confirms the anecdotal picture from Bonaparte, Bourcier, and Lesson's birds: *Aloe* has twelve species named for someone's wife, but no species at all named for anyone's husband.

The shortfall in namings for husbands is just one symptom of the regrettable exclusion of women from science, and thus from conferring species names. It's of course not remotely the most important one (and the injustice done is to the scientifically excluded women, not the nomenclaturally excluded men). It's encouraging, though, to recognize signs of progress in twentieth- and twenty-first-century naming. Wives still appear, to be sure, as in the ant-parasitic fungus *Ophiocordyceps albacongiuae.* That name (from 2018) honors Alba Congiu, wife of entomologist David Hughes; and while some might think it her bad luck that he studied parasitic fungi and not birds of paradise, the tribute is nonetheless genuine. But today, husbands have joined those wives. For example, in 1984, Angeles Alvariño named a new Antarctic siphonophore (a relative of jellyfish and corals) *Lensia eugenioi* for her husband Eugenio Leira; and in 2005 Daphne Fautin named a sea anemone *Anthopleura buddemeieri* for her husband Robert Buddemeier. Fautin made clear in a media interview that in naming *A. buddemeieri,* she wasn't implying any physical resemblance between anemone and husband. Sasanka Ranasinghe named the goblin spider *Grymeus dharmapriyai* in 2018 for her husband Prasanna Dharmapriya; but while a "goblin spider" might seem doubly pejorative, Ranasinghe chose a species with a prettily decorated and heart-shaped sternum (the plate covering the underside of the spider's thorax). Dharmapriya's goblin spider, then, is something of a living valentine card. Finally, there's the lichen *Bryoria kockiana.* This species had a less direct path from eponym to name. Its discoverers auctioned the naming rights to their new species, in a fund-raiser for a wildlife crossing on a British Columbia highway. The winning bidder was the wildlife artist Anne Hansen, who requested that the species be named for her late husband Henry Kock.

Love doesn't always follow a straight line, so some Latin names commemorate relationships that are a little more complicated. The entomologists Kelly Miller and Quentin Wheeler, in 2005, named species of slime mould beetle for their respective wives: *Agathidium amae* and *Agathidium marae* (you may remember our earlier discussion of *Agathidium,* with its species named for George Bush, Donald

Rumsfeld, and Dick Cheney). One more, *Agathidium kimberlae,* is named for Wheeler's ex-wife, "for her understanding and support of his taxonomic ways for a quarter century of marriage."[4] Wheeler explained that he'd promised, before their split, to name a species for her, and with *A. kimberlae* he followed through. The relationships recognized in Wheeler's namings were clearly described, but there's sometimes a deeper air of mystery. Consider, for example, the French botanist Raymond Hamet, who in the 1910s named three plant species for a woman named Alice LeBlanc. Alice has since been described as either a "friend" or an "intimate acquaintance" of Hamet's, and they were clearly very close. Hamet's description of *Sedum celiae* ("celiae" is an anagram of "Alice") dedicates it to her with "great affection," while his description of *Kalanchoe leblancae* calls her his "great friend" and suggests that the species name should remind her of "the languid charm of that autumn evening" when they first collected it.[5] But in naming the third species, Hamet outdid himself. That's *Kalanchoe mitejea,* described jointly by Hamet and LeBlanc—with "mitejea" being an anagram of "je t'aime" (French for "I love you"). Little is now recorded about Alice Leblanc. Perhaps they were young lovers who drifted apart, or perhaps they had a fling outside marriage; for Alice was almost certainly not the wife who is mentioned, but not named, in Hamet's eventual obituaries.

If Hamet was just a little coy about Alice LeBlanc, Ernst Haeckel didn't see much need for similar discretion. Haeckel (1834–1919) was a brilliant German polymath whose interests included philosophy, marine biology, art, and, unfortunately, attempts to use evolutionary ideas to justify racist beliefs in the superiority of Europeans over other human cultures. Haeckel discovered, described, and named thousands of species (most of them marine invertebrates) over a career studded with scientific accomplishments. His personal life was not so accomplished. Haeckel's first wife, Anna Sethe, died tragically just eighteen months after their marriage, and Haeckel was devastated. Although he remarried, that second marriage (to Agnes Huschke) seems to have held little joy for either party. There can be many signs of a marriage gone cold; but unusually, in Haeckel's case, these included the Latin names he coined. He did name at least one species (a radiolarian, a kind of tiny shelled amoeba) for Agnes, but he made his feelings clear in naming a beautiful jellyfish for Anna. He described the latter species in 1879 as *Desmonema annasethe,* and in 1899 wrote about it in a book of prints, *Art Forms in Nature:* "The species name of this extraordinary Discomedusa—one of the loveliest and most interesting of all the medusae—immortalizes the memory of Anna Sethe, the highly gifted, extremely sensitive wife of the author of this work, to whom he owes the happiest years of his life."[6]

This is lovely, or at least it would be, but for the fact that when Haeckel wrote it, he was married to Agnes. Her reaction to this public snub is unrecorded, but one assumes it couldn't have made her happy. Agnes was presumably even less happy with *Rhopilema frida,* which Haeckel named in honor of his lover Frida von Uslar-Gleichen. Frida had written to Haeckel in 1898 praising one of his books, and they struck up a correspondence that over just six years would run to more than 900 letters. At first they wrote about science, but the letters soon became personal and increasingly heated. In June of 1899 they arranged to meet. Frida was 34 or perhaps 35 years old; Haeckel was 65 (and Agnes, still his wife, was 56). He later wrote of the "shiver of desire" that came with the affair's first kiss; and meeting Frida again in March of 1900, he wrote to her: "Love . . . you are my ideal—the real ideal of a living wife . . . After I waved the last good-by at your departure this morning at 6:15, I remained another two hours in our romantic hotel!! Your 'great mad child' committed all sorts of

Frida von Uslar Gleichen's jellyfish, *Rhopilema frida*, center (Illustration by Ernst Haeckel, from *Kunstformen der Natur*, plate 88; public domain)

foolishness—washed himself yet again . . . out of your washbasin, celebrated solemn memories in each of [our] two magical rooms."[7]

But Haeckel could never quite decide what to do about Frida. He refused to leave Agnes, but also refused to stop seeing Frida, and by all accounts, he found his own indecision agonizing. The affair, if not the agony, ended in 1903 when Frida died of a morphine overdose, and Haeckel settled back into life with Agnes. He never told Agnes about Frida (although it's inconceivable that she didn't suspect); but in *Rhopilema frida,* he told the world.

Haeckel admitted to "all sorts of foolishness" over his love for Frida, but love—or at least, infatuation—has driven people to foolishness since history began. Even Linnaeus, who took botany, and naming, and himself so very seriously, once allowed what seems to have been an unrequited crush to cloud his judgment. It happened late in his life: in 1767, at age 60, he named a genus of plants related to geraniums *Monsonia,* for Lady Anne Monson. That's not the foolish part, for *Monsonia* is a perfectly sensible naming. Monson was noted for her botanical work, and although Linnaeus had never met her, he would have been familiar with her contributions to the discipline. Not only that: species of *Monsonia* are found in India and southern Africa—both places where she lived or collected, as she accompanied her husband Colonel George Monson who was stationed with the British military in Calcutta. In fact, *Monsonia's* naming would seem completely unremarkable except for a draft of a letter from Linnaeus to Monson (Anne, not George), in which he rhapsodizes in cringeworthy fashion:

> I have long been trying to smother a passion which proved unquenchable and which now has burst into flame . . . I have been fired with love for one of the fair sex, and your husband may well forgive me so long as I do no injury to his honor. Who can look at so fair a flower without falling in love with it, though in all innocence? . . . I have never seen your face, but in my sleep I often dream of you. So far as I am aware, Nature has never produced a woman who is your equal—you who are a phoenix among women. . . . But should I be so happy as to find my love for you reciprocated, then I ask but one favor of you: that I may be permitted to join with you in the procreation of one little daughter to bear witness of our love—a little *Monsonia,* through which your name would live forever in the Kingdom of Flora.[8]

It seems likely the Linnaeus never sent this letter, suggesting that the lapse in judgment it represents was short-lived. What, one wonders, could he have thought he'd accomplish in writing it? It's true that he professes—twice—that his love is innocent; and the "procreation" of a little daughter seems to be metaphorical (he's referring to the plant *Monsonia,* not propositioning the Lady Anne to bear his actual child). But if his intent was innocent, he certainly didn't hold back on the double entendres—even by the somewhat overwrought standards of eighteenth-century prose. Many *Monsonia* species have petals tinged with pink, and it's difficult to look at them now without thinking of the blush that would surely have colored the faces of both Linnaeus and Anne Monson had the draft letter been finished and delivered.

I began this chapter by suggesting that recording love in a Latin name shows scientists as humans at their best. Haeckel's and Linnaeus's stories may not be very good examples; and it's true, after all, that love sometimes drives people to foolishness or worse. So let's finish with an example that can stand as an inspiration: the snail *Aegista diversifamilia,* named in 2014 by Chih-Wei Huang and colleagues. *A. diversifamilia*'s name doesn't refer to any particular loved individual, but rather to the notion that all should be free to love, and that all love should be celebrated equally. More specifically, the name *A. diversifamilia* (meaning, in Latin, "different family") was named in support of equal rights for same-sex marriages. When Huang's new snail species needed a name, the issue of same-sex marriage was being hotly debated in his home country, Taiwan. (A Constitutional Court decision in 2017 tossed out laws recognizing only heterosexual marriage. A two-year window given for the legislature to act was largely consumed by protests and counterprotests, but in May 2019, with just a week to spare, Taiwan's legislature passed a bill legalizing same-sex marriage. That won't end the debate, but it was a victory for love.) The authors explained in a press release that the snail's hermaphroditic mating system (all individuals are both male and female) is very different from our own, and can represent the diversity of sexual orientation and relationships in the animal kingdom; its name can represent the diversity of relationships in our own species.

Like Bonaparte's bird of paradise *Diphyllodes respublica,* the name *Aegista diversifamilia* has been criticized because it's a bit of political expression coming through a channel (the scientific literature) where we like to think there isn't any. But it's a name, not a scientific conclusion; it makes a statement not about how the natural world works, but about how the human world should. Love is universal, or ought to be, and if the name of a snail can remind us of that, so be it.

15

The Indigenous Blind Spot

Eponymous Latin names have a lot to teach us about history, biology, and the culture and practice of science. Individually, they honor thousands of people who have made contributions to science or to society more broadly. But taken collectively, what they *don't* do is paint any kind of representative picture of those who deserve such honors. That's because in naming, as in so many other things, we've done a conspicuously bad job in distributing the honor with respect to human diversity. Species by the hundred, for instance, are named for Victorian British naturalists and explorers—again and again, white men, usually privileged by family background, like Darwin and Bates and Wallace and Spruce. The pattern extends beyond Victorian Britain, of course: there are species for Humboldt and Rudbeck and Cope and Buffon, and the list of white Western men could continue for hundreds of pages.

Consider, less anecdotally, a recent compilation of namings in the large plant genus *Aloe.* The authors of that compilation documented 278 eponymous namings in *Aloe,* of which fully 87 percent were for men (primarily, for white Western men). This surely reflects the long-standing exclusion of women from opportunity in science and the fact that many names were bestowed decades or even centuries ago, when that exclusion was particularly rigid. But even allowing for that, there's clearly a collective mischaracterization going on in *Aloe* naming—and by extension, in naming more generally. It's not that Darwin and Bates and the like didn't make enormous contributions to science; of course they did, and they deserve the honor of their eponymous names. It's just that nobody should imagine that they're alone in deserving eponymic honor.

The mischaracterization of human diversity in naming is particularly severe in the under-recognition of Indigenous people. Indigenous eponyms aren't completely absent: some industrious searching will turn up dozens of species named for Indigenous people. But "dozens," of course, is a tiny drop in the vast bucket of named species—albeit a tiny drop that reveals some interesting patterns.

To begin: there seem to be quite a few names that honor Indigenous *peoples* rather than Indigenous individuals. For example, the tiger beetle *Neocollyris vedda* and the goblin spider *Aprusia veddah* are named for the Vedda people of Sri Lanka, the microfossil *Cerebrosphaera ananguae* for the Anangu people of southwestern Australia, and the leafcutter bees *Hoplitis paiute, H. shoshone,* and *H. zuni* for three Indigenous peoples of the American southwest. Typically, such names recognize the peoples associated with the area inhabited by the named species. Sometimes, such namings can be poignant, as in the case of the Umpqua chub, a highly endangered fish found only in the Umpqua river of Oregon. Its scientific name, *Oregonicthys kalawatseti,* refers to the Kalawatset or Kuitsh people. The Kalawatset lived around the Umpqua river, and in the late 1700s and early 1800s came into frequent and unfortunate contact with European explorers and settlers in the area. It's a common, sad story: in addition to direct conflict with settlers, they were afflicted by European diseases and by large-scale environmental changes associated with settlers' land use. By the late 1800s their numbers were severely depleted, and although there are modern descendants, their Kalawatset language and much of their culture have been lost. Now, on the very same Oregon land-scape, the Umpqua chub and many other species are threatened by some of the same forces. The namers of *O. kalawatseti* put it this way: "Oregon once had a remarkable diversity of native peoples with more native languages than all of Europe. The Kalawatset . . . were part of this lost human diversity and [our naming can] serve to forewarn of a parallel decline in diversity of Oregon's native freshwater fishes."[1]

What about species names honoring *individual* Indigenous people? Here we find something that's curious indeed: while such names do exist, a remarkably large fraction of them refer to emperors, kings and queens, and military leaders. As a few examples, consider:

- the moth *Adaina atahualpa* and the frog *Telmatobius atahualpai* (Atahualpa, the last Incan emperor);
- the fruit fly *Drosophila ruminahuii* (Rumiñahui, an Incan general who led the last resistance to the Spanish after Atahualpa's death);
- the butterfly *Parides montezuma* (Montezuma, an Aztec emperor);
- the swordtail fish *Xiphophorus nezahualcoyotl* (Acolmiztli Nezahual-coyotl, ruler of the Alcohua city-state of Texcoco in central Mexico);
- the butterfly *Vanessa tameamea* (Kamehameha, the royal house of Hawaii, based on a dubious transliteration);
- the pelican spiders *Eriauchenius andriamanelo, E. andrianampoin-imerina, E. rafohy, E. ranavalona,* and *E. rangita* (for two kings and three queens of the precolonial Merina Kingdom of Madagascar);
- the Shawnee darter, *Etheostoma tecumsehi* (for Tecumseh, the Shaw-nee chief and warrior who fought American forces but allied with the British in the early 1800s).

It's easy to come away from this list with the impression that we've been fascinated by royalty and military leadership in Indigenous cultures. That's not surprising, as we're equally fascinated by royalty and military leadership in Western cultures. There are, of course, eponymous names for Western royalty too—Queen Victoria's water lily, *Victoria amazonica,* with its floating leaves 3 meters in diameter, is a familiar (and spectacular) example. But the window opened on human history and culture—Indigenous or otherwise—by a list of emperors and generals is a very limited one. In this light, the earth snake *Geophis juarezi* is refreshing: its name honors Benito Juárez, the first Indigenous president of Mexico. Juárez was of Zapotec heritage, and even though his presidency (he was first elected in 1861, and served until his death in 1872) wasn't without controversy, he's remembered in large part as a reformer who advanced the rights of Mexico's Indigenous peoples.

Emperors and queens and presidents don't have much to do with science (at least, not directly). I began this chapter by remarking on the frequency, among eponymous names, of those for Western naturalists such as Darwin, Bates, Cope, and Buffon. What about namings for Indigenous people who have made similar contributions to science? But if names for *any* Indigenous people are uncommon, names for those involved in the pursuit of scientific knowledge are vanishingly rare. There are probably two reasons for this. First, and most obviously, there's a long and sorry history of exclusion of Indigenous people from the scientific enterprise—indeed, from education and society more broadly—as a legacy of European colonialism. To a great degree, Indigenous people have been prevented from making the contributions that namings honor (and that's a far bigger topic than this book can address). But second, I'd argue that we've failed to acknowledge the contributions that Indigenous people *have* made. Those contributions go back a long way, but they tend not to be in our textbooks.

The era of Western exploration and colonization—especially the eighteenth and nineteenth centuries—saw tremendous progress in our scientific knowledge of Earth's biodiversity. (It also saw enormous and widespread human suffering, of course, for the Indigenous communities whose homelands were colonized.) The era's scientific progress can be accounted for, at least in large part, by two kinds of global movement: a flood of specimens returning from expeditions overseas to European and North American museums, gardens, and zoos, and a reciprocal flood of collectors and naturalists mounting, or joining, exploratory expeditions to parts of the world newly available to Western study. Some of the explorers, collectors, and naturalists involved in each flood are now obscure (such as Colonel Robert Tytler, whom we met in Chapter 10); others, like Darwin, are household names. But few—if any—of them acted alone. Expeditions and collecting trips often had Indigenous guides, field assistants, and other support workers, and the contributions these people made weren't trivial—many an expedition would have failed miserably without them. A recent compilation by John van Wyhe, for example, suggests that Alfred Russel Wallace's famous expedition to the Malay Archipelago probably involved well over 1,000 local assistants. The best Western naturalists understood that Indigenous people were an invaluable source of information about the local flora and fauna. Hunters, gatherers, and healers in particular often possess detailed knowledge not just of what species live in their region but of those species' ecology and behavior, and of where and when to find them and how to acquire specimens. Indigenous peoples' conceptual systems for organizing their knowledge of flora and fauna tend to be sophisticated and to map remarkably well to later scientific assessments of biodiversity. For all these reasons, many "discoveries" of new species by Western science wouldn't have happened without Indigenous contributions.

Indigenous contributions to science, though, are often ignored or at least underemphasized. Traditional knowledge of species, for example, often receives the dismissive-sounding label "folk taxonomy." This is part of a broader discomfort among Western scientists with Indigenous traditional knowledge—or, as perhaps we ought to call it, Indigenous science. Indigenous insights are often dismissed as anecdotal or lacking basis in formal data, although it's not uncommon for a "new" scientific discovery to be trumpeted despite its familiarity to Indigenous people who have known about it for millennia.

All this has unsurprising consequences for naming. Indigenous contributions to our understanding of Earth's biodiversity seem to have been recognized only very rarely with eponymous names. The French ornithologist François Le Vaillant named a South African cuckoo after his Khoi guide and wagon-driver Klaas, whom he praised as a brother and described in his typically purple prose as "generous Klaas, young pupil of nature, whose virtuous mind was never corrupted by our elegant institutions."[2] Or rather, he halfway named a cuckoo for Klaas. Le Vaillant, working at the close of the 1700s, was one of the last holdouts against the Linnaean system of naming. When he described a new species (and he described a lot of them), he gave it only a common name—in this case, "Klaas's cuckoo." It fell to the English zoologist James Francis Stephens, 20 years later, to make a formal redescription and confer the scientific name *Cuculus klaas* (it's now *Chrysococcyx klaas*).

Le Vaillant's is the earliest Indigenous naming I've been able to trace, but there were a few more from the colonial days of exploration and species discovery. The British ornithologist Edgar Leopold Layard named a Sri Lankan flycatcher *Butalis muttui* (now *Muscicapa muttui*) "after [his] old and attached servant Muttu, to whose patient perseverance [he owed] so many of [his] best birds."[3] Similarly, the German botanists Paul Ascherson and Georg Schweinfurth named an African member of the willow family *Homalium abdessammadii* in honor of Mohammed Abd-es-Ssammâdi, Schweinfurth's Kenyan "most faithful friend" and traveling companion (Ascherson and Schweinfurth 1880:130). (Whether Abd-es-Ssammâdi should be considered Indigenous is debatable, with Kenya's pre-European history of settlement, resettlement, movement, and colonization being complex.)

More familiarly, the Shoshone woman Sacagawea accompanied the famous Lewis and Clark expedition (1804–1806) across the northwestern United States, acting as a translator, guide, and natural historian. Among other things, after a mountain crossing that reduced the party to eating their tallow candles and their horses, Sacagawea is credited with identifying camas roots as a food source (the Latin name for camas, *Camassia quamash,* comes from Clark's recording of a Niimí'ipuutímt name for the plant). The Lewis and Clark expedition returned, after two years, with a wealth of natural history observations and specimens (including at least 94 plant species new to science). At least four species now bear Sacagawea's name: the scorpionfly *Brachypanorpa sacajawea,* the cranefly *Tipula sacajawea,* the hoverfly *Chalcosyrphus sacawajeae,* and the bitterroot *Lewisia sacajaweana.* The last is an interesting one: it's a plant belonging to a genus named for Lewis, and all 19 species in the genus have large, edible roots that were harvested by Indigenous communities in western North America. It's a good name, if not quite a perfect one, as *L. sacajaweana* is a rare local endemic (in parts of central Idaho) that Sacagawea probably never saw. She would have known the genus, though.

Namings for Sacagawea and others from the era of colonial exploration are bittersweet in a way. They honor individuals who made real contributions, but at the same time, the supporting roles in which they made those contributions mark the exclusion of Indigenous people, all over the world, from more central roles in science. Very slowly—too slowly—that exclusion is beginning to crack. That's why I'm so pleased by the tidepool's-worth of algae named for Isabella Abbott. Abbott was the first Indigenous Hawai'ian woman to earn a Ph.D. in science (in 1950), and she became the world's expert on the marine algae of the tropical Pacific. Over a long and brilliant career, she discovered and named more than 200 species of algae, wrote extensively about Hawai'ian traditional knowledge of botany and marine biology, and taught generations of students to love both subjects. She was deeply respected by her fellow phycologists ("phycology" being the scientific study of algae), who repeatedly took the opportunity to name newly discovered algal species after her. Among "her" algae are *Pyropia abbottiae, Dasya abbottiana, Udotea abbottiorum, Phydodris isabellae, Liagora izziae,* and the genera *Abbottella, Isabbottia,* and *Izziella. Izziella* is a special treat, because the first species described was *Izziella abbottae*—with Isabella ("Izzie") Abbott honored in both the genus and species names. (That name, disappointingly, is now considered a junior synonym of *Izziella orientalis.*) Sadly, many past Isabella Abbotts have seen their potential wasted due to lack of opportunity. Many more Isabella Abbotts are out there, needing only that opportunity to bring their talents to the scientific enterprise; including them can only strengthen that enterprise.

There are twin deficits, then: too few Indigenous people accepted into the fold of the scientific community, and too few acknowledgments of what Indigenous people have contributed to science nevertheless. But: we've named around a million species, and there are at least several million more to come. Those unnamed species represent opportunities, and we can take them to acknowledge the diversity of contributions to science. This means a broader diversity of eponymous names, and a deliberate effort to learn about deserving candidates for the honor. They're surely out there.

Isabella Abbott's red alga, *Izziella abbottae*

There's an important note of caution to be sounded here, though. Before we rush to name thousands of new species after Indigenous people, we should think carefully about respect for the Indigenous cultures involved. In some Indigenous communities, eponymous namings will be welcomed, seen by community members as the honor that's intended. Other Indigenous communities could have different reactions, for at least two reasons. First, those Indigenous people who supported Western expeditions may have (surely inadvertently) therefore supported the colonization that brought so much misery. The Lewis and Clark expedition, for instance, may not have had the goal of disrupting and displacing Indigenous communities in the American Northwest, but there's little doubt that the knowledge it returned contributed to that outcome. Eponymous names for Sacagawea, while intended to honor her contributions to science, might understandably be seen by some in Indigenous communities as referring instead to collaboration with Western colonialism. Second, the notion of honoring someone with an eponymous name (for a species, a mountain, a building, or anything else) may make sense in most Western cultures, but there's no reason to expect this to be a universal human practice. In western North America, for example, traditional Indigenous place-naming typically doesn't involve eponymy; rather than naming places after people, members of many cultures in the region name people after places. If people in those communities think similarly about naming species, an eponymously named species might be met with puzzlement rather than gratitude.

In other Indigenous cultures, tradition around the power of names might make eponymous naming seem disrespectful or threatening. In a group of New Zealand Māori interviewed by Judy Wiki Papa, for example, there was strong support for using the Māori language in naming new species, but not for eponymous namings. One person put it this way: with respect to ancestor's names, "that was a name that belongs to them, they have recognition amongst their people and that is where their honor should lie, amongst their people."[4] Names, for Māori, don't just identify people; they also narrate lives and genealogy, transmit knowledge, and represent the place of the named person in a connected universe. An eponymous naming, as a result, could well be seen as offensive. Among many Australian Indigenous communities, the reaction would be even stronger, because of cultural practices that forbid speaking or writing a recently deceased person's name. The length of the name-avoidance period and the strength of the practice vary across communities, but the implications for eponymous naming are obvious. Even this is too simplistic, though: within one Indigenous society we should expect as much diversity of opinion as we find in our own, and that means eponymous namings might be welcomed by some individuals but found offensive by others. All this suggests caution, and I suppose it's possible that respectful restraint is a *third* reason for the dearth of Indigenous eponyms (in addition to the exclusion and omission discussed earlier). That would paint science in a better light, but I'm skeptical; it would account for a lack of eponymous names for *some* Indigenous cultures, but not for a lack of eponymous names across *all* Indigenous cultures.

Perhaps I seem to have painted science into a corner here: to continue bestowing Western eponyms is to perpetuate a colonial attitude to our work, but to reach for Indigenous eponyms instead is to risk cultural trespass. How are we to navigate this complexity? Simply giving up on eponymous naming altogether sacrifices an opportunity to recognize people (and peoples) who've been marginalized both in science and in its recorded history. Instead, those considering an eponymous naming should talk with members of the relevant Indigenous community. That may be the particular person to be honored, of course, or descendants of that person, or community leaders or elders who can give some guidance as to whether an eponymous naming would sit well in the particular Indigenous community at issue. A good example is provided by David Seldon and Richard Leschen, who in 2012 named six new beetle species from New Zealand. Each was given a species name with Māori etymology (in this case, none of them eponymous), but only after consultation with Māori people in the areas where the beetles were discovered. This kind of consultation doesn't guarantee lack of offense, of course, because all societies include differences of opinion. More generally, in life, *nothing* guarantees that an act will offend no one. But the consultation itself is a gesture of respect, and it greatly increases the likelihood that a naming will be seen as an honor.

Isabella Abbot's algae—*Izziella abbottae* and the rest—can perhaps be seen as something of a beacon. They suggest that scientists can diversify their eponymous naming to include Indigenous people, while acting with respect toward cultures with different attitudes toward naming. Now, *Izziella* may not be everyone's idea of a charismatic species—it's a fairly small, fairly spindly, fairly obscure seaweed. But it's a small, spindly, obscure seaweed belonging to a group that Isabella Abbott loved, and a group that she made immense strides toward understanding. It's an appropriate tribute indeed. The naming of *Izziella* (even with a thousand more namings like it) isn't going to solve all the world's post-colonial problems. But small steps advance a journey too.

16

Harry Potter and the Name of the Species

Wasps have, in general, a bad reputation. So do Malfoys.

Wasps are members of the insect order Hymenoptera, along with the bees and the sawflies. It's an enormously diverse group of animals: over 150,000 species have been described and named so far, and there are at least as many more waiting to be discovered. Wasps are fascinating and often beautiful, but they tend not to be among people's favorite insects. For many, the mental image conjured by "wasp" is of being harassed at a picnic, stung while cleaning gutters, or driven from one's patio by an aggressive band of squatters. By far the majority of wasp species take no part in any of these things—but in a way, they're even less immediately lovable. Most wasps are "parasitoids": smaller insects (by and large), whose adults lay eggs on or into other living insects, and whose larvae then develop and feed inside the living bodies of their unfortunate hosts. An insect harboring a parasitoid may go about its normal business, unaware of its doom; or it may even have its behavior manipulated by the wasp larva so that it feeds more and longer, so there's plenty for the developing wasp to feed on in its turn. The host insect *is* almost certainly doomed: in a moment worthy of a horror movie, the parasitized insect normally dies when the full-grown wasp bursts from the body of the host it no longer needs. No wonder, really, that wasps have their bad reputation.

It's against this backdrop that Thomas Saunders and Darren Ward reported the discovery of a new species of parasitoid wasp, from New Zealand. Their new species belongs to the genus *Lusius,* and they jumped at the chance to name it *Lusius malfoyi.* Lucius Malfoy (with a 'c') is a character in the *Harry Potter* stories, so there's both a pun and a reference to the beloved books. There's a more interesting connection, too. Lucius is mostly a sinister character in the books, a member of the evil Lord Voldemort's "Death Eaters" in their conflicts with the forces of good. He's an interesting and complex character, though. After the First Wizarding War, Lucius claims that his allegiance to Voldemort's side was compelled by a magical spell (although it's not clear whether he's telling the truth). By the climactic Battle of Hogwarts he's been largely renounced by Voldemort and plays little obvious role, proving to be motivated more by love for his wife and child than by any higher calling to evil or to good. Saunders and Ward suggest that wasps, like Lucius Malfoy, need to be understood with nuance. Relatively few species annoy humans or cause economic harm, while many serve beneficial roles by controlling agricultural pests (among other things). All have evolved by means of natural selection, which doesn't shape either good or evil but which rewards care for an individual's offspring. Finally, as for *Lusius malfoyi:* it's a parasitoid, but because it's only ever been collected in studies that net adults from vegetation, we don't know which host insects it attacks. Like its namesake human Lucius Malfoy, the wasp *Lusius malfoyi* plays a role in the larger sweep of nature that isn't clear to us—or at least, it isn't clear yet.

Lusius malfoyi isn't the only species named after a fictional character. In fact, it's not even the only species with a *Harry Potter* name. The crab *Harryplax severus,* for example, is a double Potter reference. The genus, *Harryplax,* was named for the collector of the specimens, Harry Conley, but the namers also suggest an allusion to the more famous Harry (Potter) via Conley's "uncanny ability to collect rare and interesting creatures as if by magic."[1] The species name, *severus,* refers to Severus Snape—the wizard and Hogwarts professor whose motivations and backstory were hidden through seven novels to be revealed only at the series' climax. This, the namers suggest, is akin to the crab's having remained unnoticed and undescribed for 20 years since specimens were first collected. Also among *Lusius malfoyi's* etymological brethren are at least three different spiders named after Aragog, the giant spider raised by Hogwarts groundskeeper Rubeus Hagrid. The three are *Ochyrocera aragogue, Lycosa aragogi,* and *Aname aragog,* and fortunately, none of them grows anywhere close to the five-meter legspan of Hagrid's Aragog. Finally, and most wonderfully, a fourth spider is named *Eriovixia gryffindori,* after Godric Gryffindor, one of the four wizards who founded the Hogwarts school. Why a spider for Gryffindor? The four founders together enchanted Gryffindor's hat, which ever since has magically sorted incoming students into one of Hogwarts' four "houses." The sorting hat is described in the *Harry Potter* books only as "a pointed wizard's hat . . . patched and frayed and extremely dirty," but in the movies it's conical, gray and brown, with a bent-over tip. That's exactly what the spider looks like, too, quite possibly so it can camouflage itself as a dried and rolled-up leaf. The scientists who named *Eriovixia gryffindori* explain their choice this way: "This uniquely shaped spider derives its name from the fabulous . . . sorting hat, owned by the (fictitious) medieval wizard Godric Gryffindor . . . and stemming from the powerful imagination of Ms. J. K. Rowling, wordsmith extraordinaire. An ode from the authors, for magic lost, and found."[2] Magic found, indeed, because the *Harry Potter* books are stunningly imaginative and have drawn more children into the magic of reading than anything else that's been written in many, many years.

Harry Potter eponyms are, as you might expect, only the tip of the fictional iceberg. Among other recent examples, two genera of deep-sea worms are named for characters from George R. R. Martin's *A Song of Ice and Fire* series: *Abyssarya*, for Arya Stark (plus "abyss," because the specimens were collected more than 4 kilometers deep in the Pacific Ocean), and *Hodor*, for the stableboy Hodor. Seven wasp species take their names from the same series: *Laelius arryni, L. baratheoni, L. lannisteri, L. martelli, L. targaryeni, L. tullyi,* and *L. starki,* representing the seven noble Houses of

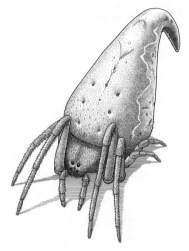

The sorting-hat spider,
Eriovixia gryffindori

Westeros. A Kenyan waterweed is named *Ledermanniella maturiniana* after Dr. Stephen Maturin, from Patrick O'Brian's *Master and Commander* series. Maturin is an excellent naturalist but an indifferent sailor who often falls out of boats, and the plant's natural habitat is frequently submerged. If your tastes run more to fantastical humor, then there's a set of wasps for you: a monograph describing 179 new species of the wasp genus *Aleiodes* included 34 names for characters in Terry Pratchett's *Discworld* novels. Since the wasps are parasitoids, and lethal to their hosts, it seems appropriate that nine are named for members of the books' Assassin's Guild: *Aleiodes tmaliaae, A. teatimei, A. selachii, A. pteppicymoni, A. prillae, A. nivori, A. flannelfooti, A. downeyi,* and *A. deathi.* Lest anyone think taxonomists only read mass-market and genre fiction, though, let's add to this list the bess-beetle *Oileus gasparilomi* (after Gaspar Ilóm, the hero of *Men of Maize* by the Nobel Prize–winning novelist Miguel Ángel Asturias) and the catfish *Ituglanis macunaima* (after Macunaíma, the hero of Mário de Andrade's modernist novel of the same name).

Of course, fictional characters needn't come from novels. There are species named for characters from comic books: among them, the weevils *Trigonopterus asterix, T. idefix,* and *T. obelix,* named for the three heroes of Goscinny and Uderzo's *Asterix* comic strips. There are dozens of species named for movie and television characters, including the trilobite *Han solo,* the wasp *Polemistus chewbacca,* and the rather sponge-like mushroom *Spongiforma squarepantsii.* A particularly interesting name is based on a 1960s-era radio character. The cyanobacterium *Calochaete cimrmanii* was named in 2014 by three Czech scientists to honor Jára Cimrman, a fictional Czech polymath. Cimrman's character was introduced on a radio series, *The Non-Alcoholic Wine Bar chez Spider,* but eventually became omnipresent in Czech culture. He was said to have lived before the First World War and to have made a remarkable variety of contributions to world literature, art, science, and sport. Among other things, he convinced Chekhov that his famous play needed three sisters, not two; he pioneered obstetrics in Switzerland; he came within seven meters of being the first human to reach the North Pole; he brought Pierre and Marie Curie the pitchblende from which they isolated radium; and he invented yogurt. His name on a bacterium seems the least science can do to help spread the word of his astonishing (if fictional) accomplishments.

Like namings for celebrities, names inspired by popular fiction often receive breathless, if short-lived, media attention. For instance, a "Top 10 New Species of 2018" list reported by the cable news network CNN featured the hump-backed Antarctic crustacean *Epimeria quasimodo* (perhaps it rings a bell?); the sorting-hat spider was covered by newspapers and television networks around the globe. It's easy to see this coverage as a suggestion that there's something novel about species names originating in fiction, but in fact it's a long-standing scientific tradition. A generation before taxonomists began naming species from *Harry Potter* and *Discworld*, they mined the cast of characters of *Hitchhiker's Guide to the Galaxy*. Two fish are good examples: a fjord-dwelling deepwater fish, *Fiordichthys slartibartfasti*, is named for the planetary designer Slartibartfast, who won awards for his work on Norway's fjords; and the blenny *Bidenichthys beeblebroxi* is named for the two-headed Zaphod Beeblebrox. A generation before *that*, it was *Lord of the Rings*. Wasps are in the picture again here, with genera for at least six of Tolkien's dwarves: *Balinia, Bofuria, Durinia, Dvalinia, Gimlia,* and *Oinia* (all named by Karl-Johan Hedqvist in a series of publications in the late 1970s). There are other species named for both Tolkien heroes and antiheroes: three weevils are *Macrostyphlus frodo, M. bilbo,* and *M. gandalf,* while *Gollum* is a genus of sharks and *Galaxias gollumoides* is perhaps the most on-the-nose naming (it's a swamp-dwelling fish with large eyes). There are species with names from Steinbeck and Nabokov, Twain and Tolstoy, Kipling and Dickens and the Brontës. The roots of the practice, though, go deeper than any of those, and nobody will be surprised to learn that it started with Linnaeus.

Linnaeus, in his *Systema Naturae*, gave names to well over 10,000 plant and animal species. Faced with the need for so many names, he cast a broad net for inspiration. He drew upon description (*Acer rubrum,* red maple) and geography (*Solidago canadensis,* Canada goldenrod). He coined eponymous names to honor botanists (*Rudbeckia,* for Olaf Rudbeck) or occasionally to insult them (*Sigesbeckia,* for Johannes Siegesbeck). He also dug deep into fiction—more specifically, classical Greek and Latin poetry and mythology. In the

Linnaean names of genera, we find allusions running the gamut from obvious to obscure. Some simply can't be missed: the clam genus *Venus* or the fish genus *Zeus*. Others, from a modern perspective, are subtle enough that most people would miss the classical references even if they find the names themselves perfectly familiar: the orchid genus *Arethusa,* the easter-lily *Amaryllis,* or (most familiar of all) *Iris,* hiding its mythological origins in gardens everywhere. (Arethusa was one of the fifty sea nymphs born to Nereus and Doris, while Amaryllis was a shepherdess in Virgil's *Eclogues;* Iris was the goddess of the rainbow and the messenger of the gods.) In the *Systema Naturae,* names like this are scattered across the tree of life, but a few branches are more heavily laden—especially the butterflies and moths (collectively, the Lepidoptera).

In the tenth edition of the *Systema Naturae,* Linnaeus named 544 species of butterflies and moths. He treated all the butterflies (193 of them) in the genus *Papilio,* placed 39 hawkmoths in the genus *Sphinx,* and put the remaining 312 miscellaneous moths in the rather unwieldy genus *Phalaena.* Still, Linnaeus was lucky to have had to deal with just 544 species of Lepidoptera. Since his time, the number of known species has ballooned to about 180,000, and at least that many more surely remain to be discovered. At any rate, Linnaeus needed a lot of names, and for reasons now lost to history he chose to rely heavily on names from classical sources. The butterflies, in particular and with only a handful of exceptions, received names from Greek mythology. Linnaeus was impressively systematic about this, using a thematically linked set of names for each of ten smaller groups of species within *Papilio* (his ten groups map crudely but noticeably onto our modern understanding of butterfly evolution, so his grasp of butterflies wasn't bad for an eighteenth-century botanist). The first two groups received names from the myths of the Trojan war—one group for the Trojan side, and the other for the Greek. Linnaeus started big, with the first two species being *Papilio priamus* (for Priam, king of Troy) and *P. hector* (for Hector, Priam's firstborn son and leading fighter on the Trojan side). Other groups use names for the Muses (the Greek goddesses who inspire literature, science, and the arts), for nymphs (goddesses of nature), and for the Argonauts (sailors of Jason's ship *Argo* on the quest for the Golden Fleece). Most of Linnaeus's butterflies have since been assigned to other genera; however, the species names are (with few exceptions) still in use. For instance, *P. priamus* is now *Ornithoptera priamus* and *P. hector* is *Pachliopta hector*.

Linnaeus named one group of species for the Danaïdes. There's some irony in those names, because these beautiful and delicate butterflies commemorate a particularly bloodthirsty tale. The Danaïdes were the 50 daughters of Danaus, king of Libya. (*Danaus* is now the genus name for the monarch, perhaps the world's most famous butterfly, but that name isn't one of Linnaeus's.) Danaus's daughters were promised as brides to the 50 sons of Aegyptus, Danaus's twin brother and king of Egypt. Danaus was pressured into arranging the marriages, and expressed his disapproval by ordering his daughters to kill their new husbands on their wedding nights. All but one of them did so. After death, they were (unfairly, one might argue) punished: ordered to fill a bathtub in which to cleanse their sins, they were given only leaky jugs and they labor at their task forever. It's quite a story to be brought to mind by the gentle flutter of a butterfly like the pale clouded yellow (*P. hyale,* now *Colias hyale,* a widespread Eurasian butterfly named for the Danaïd Hyale).

In a way, all this attention to Greek mythology meant that Linnaeus was a bit out of step with his times. In the early eighteenth century, the myths had fallen somewhat out of favor. The Enlightenment had placed the focus more on scientific achievement—so if ancient Greece was going to be in the spotlight, it would be Hippocrates and Aristotle and Archimedes, not the heroes of the Trojan war. Linnaeus, though, always marched to the beat of his own drum. He would surely have felt vindicated had he lived to see the end of the eighteenth century, when Romanticism bloomed and the Western world turned back to classical mythology as inspiration for poetry, art, and fiction. With Linnaeus's names and the Romantic movement to inspire them, nineteenth-century scientists produced a torrent of names from Greek and Latin mythology.

That torrent of classical mythology has eased, but we haven't left the Greek myths entirely behind. Consider, for example, *Myotis midastactus,* a bat species described in 2014. The genus *Myotis,* the mouse-eared bats, is a diverse group with well over 100 species distributed nearly worldwide. *M. midastactus* is known only from Bolivia, where it roosts in small groups in holes in the ground or in hollow trees. It's distinguished from its many relatives by its bright golden-yellow fur, which explains its name: *midastactus* meaning, literally, "Midas touch." We know the myth of King Midas from Ovid's *Metamorphoses,* his 12,000-line epic poem (at 234,000 words, it's just a little shorter than *Harry Potter and the Order of the Phoenix*). One day, some peasants brought Midas a drunken old man they'd captured in their fields. Midas recognized the man as Silenus, son of Dionysus (the god of wine and theater), and cared for him. In gratitude, Dionysus offered Midas one wish; and Midas, without thinking things through very carefully, wished for everything he touched to turn to gold. He quickly discovered that food and drink are part of "everything," and he had to beg Dionysus to reverse his gift. Gifts, in stories, are often double-edged swords.

The Midas-touch bat and the sorting-hat spider take their names from epic sagas written two thousand years apart. They have a lot in common, though, both drawing as they do on literature's perennial themes: heroism and kindness, betrayal and cruelty, forgiveness and redemption, enmity and love. It's fitting, then, that in the parade of species bearing names from fiction, students and wizards from *Harry Potter* march side-by-side with Greek warriors, kings, and goddesses. As they march, they tell stories about stories, and about scientists who love stories too.

17

Marjorie Courtenay-Latimer and the Fish from the Depths of Time

Near the village of Ferryhill, in northeastern England, there's a limestone quarry. At the base of the commercially valuable limestone is a rock formation called the Marl Slate: a fine-grained and finely layered rock rich in organic material and laden with fossils. The Marl Slate was laid down about 255 million years ago, when a shallow sea covered much of what is now northwestern Europe. In the early years of the nineteenth century, fossil after fossil was dug from the Marl Slate layer, and among the haul were fish unlike anything ever seen before. One of the oddest fossils was a fish with a three-lobed tail, an armor of thick and bony scales, and robust, limb-like fins. No living fish looked anything like this fossil, and in 1839 the prominent French zoologist Louis Agassiz named it *Coelacanthus granulosus*. It was clear that it represented a previously unsuspected branch of life's evolutionary tree—but a branch that, presumably, was long since extinct. Over the next century, more and more coelacanth fossils would turn up, not just from the Marl Slate but from deposits all over the world. About 80 fossil species are now known, spanning about 300 million years of Earth's history—from an Australian fossil 360 million years old to fossils from the southeastern United States that date back only about 66 million years. These most recent fossils are of a giant species, 3.5 m long, that vanished from the fossil record at the end-Cretaceous mass extinction along with the dinosaurs and thousands of other unfortunate lineages.

There was nothing surprising, in 1839, about the discovery of an extinct species; but just 50 years earlier, there would have been. When fossils started getting serious scientific attention in the seventeenth and eighteenth centuries, they were both mysterious and unsettling. What were they, and how did they get into rocks? The realization that fossils are preserved remnants of organisms that lived and died in the past raised as many new questions as it settled. When did these organisms live? Why could we find fossils of clams and fish on mountaintops, and of tropical species in temperate England (indeed, eventually, in Alaska and Antarctica)? And what were we to make of fossils that didn't resemble any currently living species?

Unfamiliar fossils posed the biggest problem of all. In those pre-evolutionary days, the possibility of a species that had lived once, but lived no more, was deeply disturbing. If species were divinely created (as was near universally assumed), what purpose could the Creator have had in breathing life into a species, and then taking it away again? And since the ensemble of created species seemed to have some kind of organization (similarities among species had for millennia led naturalists to sort species into schemes like the medieval Great Chain of Being), then surely it was abhorrent for extinction to create gaps? And finally, if species could go extinct, wouldn't life on Earth be gradually thinned out until in the end no life remained? All these troubling questions led most naturalists to interpret unfamiliar fossils as evidence that the corresponding creature still lived somewhere in an unexplored part of the world: dinosaurs in the depths of the African jungles, ammonites in the depths of the oceanic basins. It took another prominent French zoologist, Georges Cuvier, to establish beyond doubt that some fossils represent species that are no longer living. Cuvier made an inarguable case for extinction in a 1796 paper on fossil elephants: in a nutshell, he showed that mammoth and mastodon fossils were clearly different from either of the two species of living elephant, and pointed out that it was simply impossible for an elephant species to live anywhere on Earth without being noticed by humans. Elephants are, after all, large and unsubtle creatures.

Cuvier thus laid the groundwork for Agassiz to describe fossil coelacanths as extinct species from deep in Earth's past. If coelacanths had flourished for 300 million years but then were lost along with the dinosaurs (although Agassiz didn't know the precise timeline), this was perhaps disappointing but no surprise. Nobody went looking for living coelacanths, just as nobody (well, nobody serious) any longer goes looking for living dinosaurs. As a result, when a coelacanth turned up among the catch on a South African fishing boat, in 1938, it was the zoological surprise of the century.

Three people were key to the discovery of the living coelacanth: a fishing-boat captain, a chemist with a passion for fishes, and a young museum curator. The first caught the fish; the second recognized and named it; and the third was the most important of all. The living species now bears the name *Latimeria chalumnae* in her honor.

Marjorie Courtenay-Latimer was born in 1907 in East London, South Africa. As a child she was fascinated by birds: she declared at the age of 11 that she would one day write a book about them, and in the meanwhile she built a collection of feathers and eggs. She was also fascinated by the lighthouse on Bird Island, which was just visible some evenings from her bedroom window. As a young woman she was briefly engaged to a man who didn't approve of her "madness in collecting plants and climbing trees after birds."[1] He gave her an ultimatum: nature, or him; and she doesn't seem to have deliberated for long. Her great ambition was to work in a museum, but few museums were hiring and fewer still were hiring women, so in 1931 she decided to pursue nurse's training. Just weeks before that course was to start, however, a naturalist friend invited her to apply for the position of curator of a new museum under construction in East London. In an interview, she impressed the museum's board with her knowledge of the South African clawed frog, and she was offered the position. She wasn't, at first, curator of very much: the museum's entire collection (she reported) consisted of six beetle-infested stuffed birds, a box of stone fragments dubiously claimed to be prehistoric tools, a six-legged piglet preserved in a bottle, a dozen prints of East London cityscapes, and another dozen of the Xhosa wars. She burned the birds, discarded the stone fragments, and started to build the museum's collection with some more convincing stone tools from her own collection and a dodo's egg from her aunt's. (Whether the egg is really from a dodo remains unclear, almost 90 years later. If it is, it's the only intact dodo's egg in existence.)

Over the ensuing years she collected everything she could get her hands on, and the museum's holdings grew. She also built a network of collaborators, two of whom would be especially important to the coelacanth story. She met the first in 1933: James Leonard Brierly Smith, a professor at Rhodes University in nearby Grahamstown. Although Smith's training and teaching were in chemistry, his avocation was fishing and the biology of fishes, and he offered to identify Courtenay-Latimer's fish collections for the museum. He could never have expected where *that* would lead. She met the second three years later, when she finally made it to Bird Island: Henrik Goosen, the captain of the fishing trawler *Nerine*. After years of entreaties, she had finally secured permits to visit and collect on Bird Island, and she spent six weeks there gathering specimens of birds, plants, shells, seaweeds, and fish—fifteen large packing cases of them. Goosen's trawler called regularly at Bird Island to catch

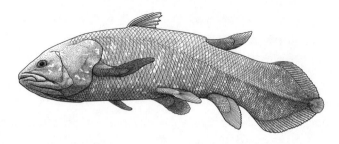

The coelacanth, *Latimeria chalumnaei*

rabbits, so the crew could have a little variety in their otherwise fish-dominated meals. He met Courtenay-Latimer there, and offered to transport her packing cases to East London. Even more usefully, he offered to save fish and other marine organisms from his trawls for the museum. He made it a habit to set aside interesting creatures—sharks, starfish, whatever seemed a bit different—and to call Courtenay-Latimer, when the *Nerine* docked in East London, to come and collect them.

On December 22, 1938, one such call came in. Courtenay-Latimer was busy preparing a fossil for display (it was a "therapsid" mammal-like reptile, *Kannemeyeria wilsoni,* that she and several colleagues had excavated from a nearby farm; it wound up being named for Eric Wilson, who did most of the digging). She was reluctant to leave her task, but decided she should at least go down to the docks to have a look and to give her Christmas greetings to the crew. She arrived to find a heap of fish on the *Nerine*'s deck, and began to sort through it. Most specimens she discarded as familiar, but underneath she saw something striking: "I picked away the layers of slime to reveal the most beautiful fish I had ever seen. It was five feet [1.5 m] long, a pale mauvy blue with faint flecks of whitish spots [and] an iridescent silver-blue-green sheen all over. . . . It had four limb-like fins and a strange little puppy-dog tail. It was such a beautiful fish . . . but

I didn't know what it was."[2] She and an assistant bundled the fish into a grain sack and, after some work to convince the driver, into the trunk of a taxi. Back at the museum, she hunted through manuals of marine fish but could find nothing remotely resembling her specimen. The museum's chairman (and collector of the dubious stone "implements") dismissed it as nothing but a common rock cod, but Courtenay-Latimer was convinced she'd come across something very special.

What to do with a 1.5-meter fish that needed study but would soon start to decompose? The small museum didn't have a refrigerator big enough to hold the fish, or enough formalin to preserve it. In East London, only the mortuary and a cold-storage locker had enough refrigerated space, and neither would agree to store the fish, so her last resort was a local taxidermist. She managed to scrounge about a liter of formalin, and they wrapped the fish in formalin-soaked newspapers and a bedsheet, but that wasn't enough to preserve more than the skin. After five days, with the fish beginning to smell and seep oil, she gave the taxidermist the go-ahead to skin and mount it. Reluctantly, they had to discard the rotting flesh and viscera. If the specimen really was as exciting as Courtenay-Latimer suspected, this wasn't the way it should be preserved for study; but there just wasn't any choice.

In the meantime, Courtenay-Latimer had written to J.L.B. Smith for help in identifying her peculiar fish. Unfortunately, Smith was away on holiday, and didn't receive the letter until January 3 (well after the specimen had been skinned for mounting). When Smith opened the letter and saw the rough sketch Courtenay-Latimer had made of the fish, at first he was stumped. He knew more about South Africa's marine fish than anyone alive, and this specimen made no sense. Later, in a popular book about the discovery, he described his reaction this way: "I stared and stared, at first in puzzlement. . . . I did not know of any fish . . . like that; it looked more like a lizard. And then a bomb seemed to burst in my brain, and . . . I was looking at a series of fishy creatures that flashed up as on a screen . . . fishes that had lived in dim past ages gone, and of which only fragmentary remains in the rocks are known."[3] Smith could hardly believe it, but he thought he knew what the sketch showed: a coelacanth. He'd never seen a fossil specimen, but he'd read papers describing them, and Courtenay-Latimer's fish had to be a match—but he doubted his own judgment all the same. It was mid-February before Smith could extract himself from his duties teaching chemistry and travel to East London to see the specimen first-hand. When he did, all doubt vanished: "That first sight hit me like a white-hot blast . . . it was a true Coelacanth."[4] At that moment Smith and Courtenay-Latimer experienced what's probably the most exciting thing in the life of any scientist: they knew something, right then, about the world that nobody else on Earth knew—something that nobody else in human history had ever known. It's an exhilarating feeling, even when the new thing you know is fairly minor. For Smith and Courtenay-Latimer, the only people who knew that living coelacanths existed, it must have been indescribable.

Smith's next task was to describe and name the species. He did both in a very brief paper (just a page and a half long) published in early 1939. It opened with a proverb originating in the natural history of Pliny the Elder: *Ex Africa semper aliquid novi,* or "From Africa, there is always something new." The rest of the paper was the bombshell, explaining that the "something new" was a living fish of a lineage last recorded from fossils 66 million years old. At the paper's end, he gave the fish its Latin name, *Latimeria chalumnae.* The genus name honors Marjorie Courtenay-Latimer, of course, while the species name records the location of the catch in the estuary of the Chalumna River. There was a lot more to learn from the coelacanth specimen, even in its taxidermied state, and Smith borrowed the fish and got in the habit of working with it, or writing about it, every day from three to six in the morning and again late into the evening (he was still teaching chemistry full-time). The result was a series of longer coelacanth papers, including a 150-page monograph outlining the fish's anatomy in great detail. These publications made Smith's reputation as an ichthyologist, and in 1945 he was finally able to resign his chemistry position and take up a new one as a research professor in Rhodes University's brand-new Department of Ichthyology.

Before publishing the first coelacanth paper, Smith wrote to Courtenay-Latimer explaining his intention to name it after her. She suggested it would be better named for Goosen, whose ship had caught the fish and who had brought it to her for the museum. Without him, she argued, there would have been no coelacanth to name. She had a point, and had Smith named the coelacanth for Goosen he would have been following a familiar and long-standing taxonomic tradition. Thousands of species are eponymously named for the person who collected the first specimen: sometimes a scientist, sometimes an amateur naturalist, sometimes a professional collector. Take, for example the Australian snail species *Larina strangei, Mychama strangei, Neotrigonia strangei, Scintilla strangei, Signepupina strangei,* and *Velepalaina strangei.* Each of these names honors the species' first collector: the Victorian professional naturalist and collector Frederick Strange. Had Smith assented, and the coelacanth been dubbed *Goosenia* instead of *Latimeria,* no one would have thought it odd. But Smith was adamant that the name should recognize Courtenay-Latimer, telling her that while Goosen might have captained the boat that caught the coelacanth, "you ultimately saved it for science."[5]

Smith may well have seen more in Courtenay-Latimer than she was willing to see in herself. She second-guessed her decision to mount the fish and discard the soft tissues, saying late in her life, "I knew [the loss of the soft tissues] was all my fault, and I have had to suffer that ever since."[6] There was criticism of her decision from others, as well. The English paleontologist Arthur Smith Woodward sniped, in an otherwise laudatory review of Smith's work, that "when the specimen was sent to the East London Museum its scientific value was not appreciated, and it was entrusted to a taxidermist."[7] This was inaccurate and grossly unfair, and Smith knew it. As he put it in one paper, "Several letters . . . have contained very harsh criticism about the loss of the carcass of this fish. . . . It was the energy and determination of Latimer that saved so much and scientific workers have good cause to be grateful. The genus *Latimeria* stands as my tribute."[8] Courtenay-Latimer richly deserved that tribute, not only for recognizing the coelacanth specimen as special and laboring to preserve as much as she could, but more broadly for building the East London Museum's collection and for assembling a network of supporters and collaborators that let her accomplish far more than she could have alone. There could be no better way to recognize all this than having Courtenay-Latimer's name attached, forever, to the most sensational new-species discovery of the twentieth century.

J.L.B. Smith has received his own eponymous tributes. His name is attached to (among other species) the Maputo conger eel, *Bathymyrus smithi*. *B. smithi* was named in 1968, the year of Smith's death, by Peter Castle—a young scientist Smith had recruited to the Department of Ichthyology he'd founded. Perhaps Smith's conger eel isn't a species quite as charismatic or as newsworthy as the coelacanth; but it belongs to a small genus of deep-water eels that has its own charm to the fish aficionado. Smith would certainly have been pleased.

But there's one gap in the story. No fish (or other species) seems to have been named for Henrik Goosen, although Marjorie Courtenay-

Latimer wanted there to be. J.L.B. Smith alone described and named over 375 new fish species but, unaccountably, never recognized Goosen with one of them. Nor, apparently, has any other ichthyologist since. Goosen's role in the coelacanth's discovery hasn't exactly been forgotten—he's mentioned at least in passing in most retellings of the story. But his keen interest and long-standing cooperation with Latimer-Smith's efforts to build her museum and to document the natural history of southern Africa deserve more. Fortunately, there are plenty of other fish in the sea—and many of them still need names.

18

Names for Sale

Madidi National Park, in northwestern Bolivia, is one of the most species-rich corners of our planet. That's partly because the park, in its 19,000 square kilometers, includes habitats ranging from lowland rain forests all the way through Andean alpine glaciers; but it's also because here, in the southwestern reaches of the Amazon basin, local biodiversity is simply astonishing. The park is home, for instance, to well over 1,000 species of birds—a tenth of the world's total—in an area smaller than Vermont or Wales. Madidi is also a place where the biodiversity is poorly known: there are likely undiscovered birds flitting through its forests, and there are certainly undiscovered plants and insects and spiders (probably thousands of them). New species pop up from Madidi with regularity. Few of them make the evening news—but a small golden-furred titi monkey named *Callicebus aureipalatii* did.

Callicebus aureipalatii was described and named in 2006, from specimens collected along the Rio Tuichi and Rio Hondo at the eastern, lowland, edge of the park. Here the forests are open, dominated by palms and cut by rivers; and here the titi monkeys look and behave differently from the monkeys elsewhere in the park. The paper describing this new discovery was unremarkable in most ways. It reported the locations where the new species was sighted, and where type specimens were collected; it described the appearance, morphology, and behavior of the species; it made estimates of its population size and discussed the need for its conservation. It also assigned the monkey its scientific name, and on the surface, that's unremarkable too: *Callicebus* is the genus of titi monkeys to which the new species belongs, and *aureipalatii* is the kind of Latinesque name we're all familiar with. But upon more careful examination, *aureipalatii* might seem a bit odd. It's based on two Latin words: *aureus* ("golden") and *palatium* ("palace"). "Golden" makes sense, as a reference to the monkey's color; but why "palace"?

It turns out that the "palace" in *Callicebus aureipalatii* has an unusual origin. The scientists who discovered this new primate held an auction for the right to choose its name—and the winning bid came from an online casino, GoldenPalace.com. The name *aureipalatii* is simply a Latin translation of the casino's name (with the terminal 'i' affixed to specify that the name is eponymously constructed). *That's* why the Golden Palace monkey's naming made the news: not, or at least mostly not, because of the scientific importance of species discovery or because of the monkey's undeniable charisma. It's a story just right for rounding out an evening newscast, which is presumably why the casino joined the bidding in the first place.

What are we to make of this? It might seem bizarre, a disconcerting and brazenly commercial distortion of normal scientific practice. The involvement of an online casino—one known for publicity stunts and controversy—certainly makes it easy to get that impression. So does the startling size of the winning bid—$650,000 (U.S.). But in fact, the Golden Palace monkey wasn't the first to have its name sold, or the last. Among scientists, the practice certainly has its detractors; but it has plenty of supporters too. Those supporters point out that sales of naming rights have the potential to do a lot of good things for science and for the natural world. The $650,000 fee to name the Golden Palace monkey *Callicebus aureipalatii,* for example, was designated for conservation in Madidi National Park. In particular, it was used to hire local people as park rangers. That accomplishes two things at once: it connects the area's residents to the park, which becomes an opportunity rather than a burden in their lives, and it provides enforcement of the park's protected status. The money wasn't the only benefit, though. The auction itself drew attention to the need for conservation in Madidi (and elsewhere) and also to the important work of identifying and describing new species. Attention for species discovery is especially valuable because it's so rare. Conservation has David Attenborough and PBS's *Nature* and units in school curricula everywhere; but species discovery has few public champions. As with celebrity namings, a high-profile auction with a controversial winner gets people from all walks of life noticing, and thinking about, our efforts to document and understand Earth's biodiversity.

Callicebus aureipalatii has been by far the highest-profile pay-for-naming arrangement, but there have been others—more, quite likely, than most people would suspect. The best-established naming program is BIOPAT (Patenschaften für biologische Vielfalt, or "Sponsorships for Biodiversity"). BIOPAT operates under the aegis of a consortium of German institutions, including the Bavarian State Collection of Zoology, the Alexander Koenig Research Museum, and the Senckenberg Center for Biodiversity Research. Since its launch in 1999, it has arranged the naming of 166 species, with each "sponsorship" representing a donation of at least 2,600 euros (about $3,000 U.S.).

The vast majority of BIOPAT names have been eponymous, but they include a lot of variety. Quite a few donors have chosen to have "their" species named after themselves—thus (for example) the frogs *Boophis fayi* and *Phyllonastes ritarasquinae* and the lizards *Enyalioides sophiarothschildae, Enyalioides rudolfarndti,* and *Paroedura hordiesi.* In each case, the scientific paper describing and naming the new species has included an acknowledgment of the donor's role; for instance, the

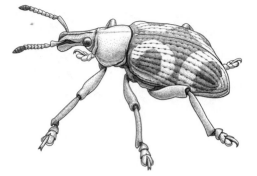

Stan Vlasimsky's weevil, *Eupholus vlasimskii*

namers of *Boophis fayi* specified that "the specific name is a patronym for Andreas Norbert Fay (Zurich, Switzerland) in recognition of his support of research and nature conservation through the BIOPAT initiative."[1]

Other donors have dedicated names to family members. The science fiction author Alan Dean Foster had a Bolivian frog named *Hyla joannae* for his wife (although in a blow for those who prefer their Latin names pronounceable, the species was later reassigned to the genus *Dendropsophus;* fortunately, it's still *joannae*). A Colombian tetra (fish) was a birthday present: *Hyphessobrycon klaus-anni* was named for Klaus-Peter Lang's parents, Klaus and Anni, on the occasion of Anni's 80th birthday. Stan Vlasimsky took family naming one step further, sponsoring an orchid for his wife Lezlie (*Epidendrum lezlieae*), and two frogs, a butterfly, and a lizard for his children Claudia, Liam, Magdeline, and Caiden (*Dendrobates claudiae, Boophis liami, Plutodes magdelinae,* and *Euspondylus caidenii*). Not to be left out, Vlasimsky also sponsored a rather spectacular New Guinean weevil, *Eupholus vlasimskii,* so the entire family could be (etymologically) together. BIOPAT namings don't end with sponsors' families, though. An orchid named for the former Soviet leader Mikhail Gorbachev

(*Maxillaria gorbatschowii*) was dedicated by a friend for his 70th birthday, while a fan sponsored another orchid for the American pop singer Anastacia (*Polystachya anastacialynae*). Several corporations have sponsored names, too, including the Danish heating- and cooling-equipment company Danfoss (the mouse lemur *Microcebus danfossi*) and the German internet-services company Pop-Interactive (the frog *Boophis popi*). These corporate namings probably reflect the influence of the splash made by the Golden Palace monkey, although neither attracted anything close to the same media attention.

BIOPAT's names have collectively raised about 582,000 euros. Coincidentally, this is just about equal to the $650,000 U.S. it took to give us *Callicebus aureipalatii*. That BIOPAT needed 166 species over 20 years to equal the haul from one Golden Palace monkey probably reflects a couple of things: that few species are as charismatic as a golden-furred tropical monkey, and that BIOPAT has taken some care not to be too conspicuously commercial. Nonetheless, 582,000 euros can do (and has done) a lot of good. BIOPAT revenues are split equally between the participating research institutes and a granting program to support biodiversity research and conservation. The money flowing to research institutes supports species discovery and taxonomic research for which funding is otherwise perpetually tight. The granting program has funded a wide variety of small projects worldwide: biological inventories of new protected areas, or areas proposed for protection; training of park rangers and government officials in biodiversity and ecology; collecting expeditions to areas likely to hold undiscovered flora and fauna; and much more.

The use of naming-rights revenue to support research and conservation seems universal, or at least close to it. An auction of naming rights for ten fishes from the western Pacific, held in 2007 by Conservation International and hosted by Prince Albert II of Monaco, raised about $2 million for conservation and educational programming in Indonesia. The star of the auction was a walking shark, now known as *Hemiscyllium galei.* That species raised $500,000 (from Janie Gale, who chose *galei* for her husband Jeffrey), which has funded patrols of a marine protected area around Wayag Island in Indonesian West Papua. The area is now a nursery for shark breeding and an economic engine for the local people. After severe budget cuts in the mid-2000s, the Scripps Institution of Oceanography turned to naming-rights sales to support and expand its specimen collections. A variety of marine invertebrates have been named, including the tropical feather-duster worm *Echinofabricia goodhartzorum,* for which Jeff Goodhartz, a high-school mathematics teacher, paid $5,000. On a smaller scale, the naming of the lichen *Bryoria kockiana* was a one-off fund-raiser for a wildlife crossing on a British Columbia highway (the winning bid was from the wildlife artist Anne Hansen, who chose a name for her late husband, Henry Kock).

Scientists who oppose the sale of naming rights (and there are plenty of them) are usually objecting to what they see as the commercialization of science. They may worry that the money sets up unintended incentives (for instance, to declare too many apparent novelties "new species" so that their naming rights can be sold). More often, though, they just have a philosophical objection to letting nature and capitalism overlap: they hold that species (and thus their names) aren't ours to sell, or they consider money for naming to taint the sanctity of science. These would be powerful objections if taxonomists were selling naming rights for personal gain. However, that kind of naming sale seems to be vanishingly rare, if it happens at all (or at least, any taxonomist who's tried it has managed to fly completely under the radar). This is actually interesting, and perhaps surprising. It isn't that there's any technical or legal obstacle to naming species for personal enrichment; nothing in the nomenclatural Codes forbids it. So one of two things must be true. Perhaps taxonomists are universally unwilling to sell names for personal profit—either on ethical grounds, or for fear of their colleagues' opprobrium. Alternatively, perhaps those who buy names are motivated at least as much by the good their money is doing (or can be seen to be doing) as they are by the name itself. Consider Jeff Goodhartz and his feather-duster worm: "It really jazzes me up," he said. "I haven't earned this like a scientist; [but] if it helps Scripps, how can it be bad?"[2] So, the selflessness of scientists, or the philanthropic instinct of donors: either is profoundly reassuring.

Of course, there are other opportunities for mischief in the sale of naming rights. What if someone wanted to pay to name something unattractive after an enemy? The BIOPAT program faced this problem early on, when a prospective customer wanted to name an insect he saw as ugly for his mother-in-law. BIOPAT declined the offer. Or what if this wasn't a "problem," but more of an opportunity? In 2014, Ph.D. student Dominic Evangelista offered naming rights for a cockroach, in an auction he dubbed "Vengeful Taxonomy." With his tongue partly in cheek—but only partly—he wrote, "We've recently discovered a new species of cockroach in the genus *Xestoblatta*. It's dirty, it's ugly, it's smelly, and it needs a name. . . . Most people have negative feelings about cockroaches, so why not name one out of spite, scorn, or revenge? Got a cheating ex-boyfriend? Hate your boss? Maybe you're just tired of hearing news about certain celebrities— *Xestoblatta justinbieberii,* perhaps? You get the idea."[3]

Evangelista was surprised that his vengeful auction, despite generating some publicity, didn't bring in many offers. The winning bidder was entomologist May Berenbaum, who requested that the new cockroach be named after herself. And so it was—as *Xestoblatta berenbaumae*— with no vengefulness sought or accomplished. Berenbaum's donation

funded Evangelista's research in Guyana, which asked whether dry sa-
vannas could be barriers to species movement and thus help spur the
evolution of new species—not just in cockroaches, but across South
America's flora and fauna. Cockroaches, just like their more charismatic
cousins, have important secrets to tell us.

So should we decry naming-rights auctions as dangerous, and as
crassly commercial and vaguely sordid distortions of scientific ideals?
Or should we welcome them as a clever tool to raise awareness of, and
money for, species discovery (and conservation)? It's an interesting
question, but in an important way it's the wrong one. The *right* question is this: how is it that the science of species discovery is so poorly
funded that its practitioners find it easier to auction off a name than
to secure grant funding to conduct their research?

We live in a world where new epidemic diseases spread, transmit-
ted by insects (or ticks, or worms)—and we don't even know how
many species of insects (or ticks, or worms) there are that might serve
as vectors. We live in a world where climate change due to emitted
carbon dioxide is an existential threat, and where green plants and
algae are by far the most important way that carbon dioxide is re-
moved from our atmosphere—and we don't even know how many
species of plants and algae there are. Our planet's biodiversity holds
the key to the discovery of new drugs, the breeding of agricultural
crops to resist pests without chemical application, and a whole lot
more besides—and that biodiversity is still, unbelievably, mostly unknown to us. To be sure, simply having species counts won't solve
many of the world's problems; but those solutions have to be built on
something, and a reasonably complete accounting of Earth's biota is a
foundation that we can't afford to neglect. Species discovery isn't
widely seen as glamorous science, like the space program or biomedi-
cal research—but it should be. It also isn't widely seen as worthy of

public funding. Museums and taxonomic research worldwide are chronically, and increasingly, underfunded (with occasional tragic results, as in the recent loss of the Brazilian National Museum to fire). The money flowing from naming-rights sales is a poor substitute for proper societal attention to the important business of documenting Earth's biodiversity; but it is, of course, better than nothing.

What would it take for us to finish the job of species discovery—to document and name every living species on Earth? It's a big job, but it's not an impractically large one. It would take a global investment in training more taxonomists, in creating university, museum, or other jobs for them, and in funding their research and housing the collections that resulted. The first serious calls for an exhaustive inventory of Earth's biodiversity came in the 1980s, when E. O. Wilson estimated that if there were 10 million species yet to be discovered, finishing the job would take the full careers of about 25,000 taxonomists. If that sounds like an implausibly large workforce, consider that in the aerospace industry, Boeing alone employs more than 45,000 engineers and nearly 100,000 other people.[4] The project nearly got off the ground in the early 2000s, following a dinner in San Francisco. Nathan Myhrvold, who had recently left the position of chief technology officer at Microsoft, was looking for projects that needed the kind of funding he and his embarrassingly rich colleagues could provide. One suggestion was Wilson's species inventory. In 2001, the "All Species Foundation" was launched, with the goal of funding an Earth inventory to be completed in 25 years at a cost estimated somewhere between three billion and twenty billion dollars. The foundation had offices, and staff, and an initial grant; but in 2002 the dot-com stock market bubble burst. The resulting crash wiped out five *trillion* dollars in paper wealth, and the era of easy money that could fund an Earth inventory, or anything else of that ambition, seemed to be over. The All Species Foundation and its efforts were suspended.

Could we begin again, and finally complete the Earth inventory? A recent study by Fernando Carbayo and Antonio Marques estimated that for animals, the job would cost about $260 billion. That estimate is much higher than those of the All Species Foundation, but it sits on a firmer foundation, with a careful attempt to price all the scientific infrastructure it would take to get the job done. Still, it's probably too low. Carbayo and Marques assumed there are 5.4 million yet-to-be-described animal species, but there may well be many more; and they didn't estimate similar costs for plants, fungi, and microbes. If we triple the estimate and round up a bit, we can guesstimate around $800 billion to discover, document, and name all of Earth's living species. That's a lot of money, to be sure; three times the cost of the Apollo program and over four times the cost of the International Space Station. It's not undoable, though. Because it would involve training new taxonomists and building new facilities, it couldn't be done overnight, so let's imagine a 20-year crash program at $40 billion per year. Forty billion dollars is less than half what the world spends on coffee, less than a quarter of what we spend shopping at Amazon, or less than 2.5 percent of global military spending. In other words: a complete inventory of Earth's biodiversity would be easily achievable, by a society that chose to achieve it. So far, we have not made that choice.

The very fact that names are for sale emerges from something of a tension between taxonomists who are convinced of the importance of species discovery and governments (and by extension the societies that elect them) that are not. As long as this tension persists—and there's little evidence that it will dissipate soon—naming-rights auctions are probably here to stay. Whether we see them as crassly commercial or as clever ways to recruit supporters to an important enterprise is up to us. With just a little frisson of nervousness, I'll make the latter choice.

19

A Fly for Mabel Alexander

It's only a fly, some might say. It's quite a handsome fly, though: metallic blue, with small bright orange antennae, enormous eyes, and a muscular thorax that suggests strong, agile flight. It's one of about 6,000 species in the family Syrphidae, the flower flies, and like its relatives, you'd see it hovering above the flowers it visits for nectar. Our fly was discovered and named in 1999 by Chris Thompson, an entomologist at the Smithsonian Institution, who found it in a box of specimens in the collection of the University of São Paulo. Its name is *Cepa margarita,* and it has (as you'll expect) a story to tell—although it's not so much the story of a fly as it is the story of its eponym, Mabel "Margarita" Alexander.

But Mabel's isn't one story; it's two. They're both true; or maybe neither is. And taken together, Mabel's two stories have something bigger to say.

Mabel Marguerite Alexander (née Miller) was born July 29, 1894, near Albany, New York. As a young woman she took secretarial training, and found work as a secretary at the Illinois Natural History Survey. The Survey was—and still is today—one of the largest state-operated research institutes in the United States, with a mission to investigate the diversity of animal and plant life in the state of Illinois. It maintains large and important museum collections, among them an outstanding insect collection (today housing about 7 million spec-

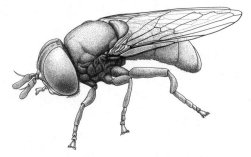

Mabel Alexander's flower fly, *Cepa margarita*

imens from around the world). Where the insects are, entomologists will come, and while she was working at the Survey, Mabel met Charles Paul "Alex" Alexander.

Alex Alexander grew up only about 50 kilometers northwest of Mabel's hometown, in Gloversville, and perhaps that connection gave them a way to break the ice when they met at the Survey in Champaign, Illinois. Alex began his career in natural history with an interest in birds, publishing his first (brief) paper in 1903 when he was just 13 years old. By 1910, though, he had discovered crane flies, publishing a paper on the crane fly fauna of Fulton County, New York. Crane flies are extremely common insects, with larvae living in freshwater, wet soil, or decaying organic matter and short-lived adults that fly freely and are often mistaken for oversized mosquitoes. Alex completed his graduate work on crane flies at Cornell University in 1917, and for the rest of his career that group of flies was his academic passion. That's why he was at the Illinois Survey—to see and work with the crane flies in its collection—so his academic passion brought him to his life's other passion: Mabel. They were married in November of 1917.

Mabel spent the next 62 years (until her death in 1979) with Alex. Over those six decades, she supported Alex's work on crane flies, becoming at first his full-time secretary and field assistant and later, much more than that. This seems to fit a once-familiar story: the wife as helpmate, devoting herself to the support of her husband's career. Perhaps that's indeed Mabel's story; but perhaps it isn't, and which story we should tell about Mabel is a question well worth asking. Before we ask it, though—before we decide on Mabel's story—let's set the stage by telling Alex's. It's easiest and most familiar to frame this rather conventionally, as a portrait of Alex as a scientist and of "his" scientific work. And that's what we'll do, for now; but of course deciding on Mabel's story has implications for Alex's too.

When Alex became interested in the crane flies as an undergraduate, something on the order of 1,500 species were known worldwide, but it was clear that this fell far short of the real total. It wasn't just a matter of the unknown wilds of the tropical rainforests, either: Alex's own collections from eastern New York were turning out to include many undescribed species. Alex began to systematically collect crane flies, looking for new species and describing them as he found them. This work continued after his marriage to Mabel, through his 37 years as a faculty member at the University of Massachusetts, and through another 20 years of "retirement" in which he used his freedom from teaching and administration to work even harder on his beloved crane flies. He published more than 1,000 papers on crane flies, and by the time the last one appeared he had named over 11,000—yes, *eleven thousand*—new species.

Let's pause for a moment to consider the magnitude of this accomplishment. Alex Alexander described, on average, more than three new crane fly species a week, and kept up that pace for seven decades (six of them with Mabel). He didn't collect all those new species himself, of course—he received specimens from collectors all over the world—but he nonetheless spent many summers on the road collecting crane flies, and other insects, from Nova Scotia to Alaska to

California. Each newly discovered species needed careful study, including comparison with every *other* known crane fly to ensure that it really was new. Once Alex was sure he had a new species, he needed to assign it to a genus (that is, decide which other species were likely its nearest relatives), and then write a detailed description of its morphology that would justify both its status as a new species and its placement among others. That "morphology" isn't just obvious things like wing shape and color, though; in crane flies as in many insects, the exact placement of tiny setae (hairs) and the detailed structure of the genitalia are critical, and these require careful dissection and microscopic examination. Finally, of course, each new species needed a name. Alex did all this, remember, over 11,000 times; and nearly 40 years after his death, those 11,000 new species of "his" *still* represent well over two-thirds of the world's known crane flies. You simply can't study crane flies, anywhere in the world, without reading his papers or identifying species that he named.

Alex Alexander doesn't, surprisingly, hold the record for publishing the most species descriptions—that record is held instead by one Francis Walker, an English entomologist who died in 1874 with 23,506 namings as his legacy. It's not a record, or a legacy, to be all that proud of. Walker's work was sloppy, and many thousands of his newly named "species" are now understood not to have been new species at all. They were species that had already been described and named by someone else (or in quite a few cases, by Walker himself), and Walker's names are thus only junior synonyms. One is not supposed to speak ill of the dead, but the opening sentence of Walker's obituary in a British entomology journal read: "More than twenty years too late for his scientific reputation, and after having done an amount of injury to entomology almost inconceivable in its immensity, Francis Walker has passed from among us."[1]

Alex's work was nothing at all like Walker's. Yes, in a few cases something described by Alex as a new species has turned out not to be, but that happens to *everyone* who describes and names more than a few species. Michael Ohl has estimated Alex's "synonymy rate" at something between three and five percent, which is admirably low (Walker's, for comparison, was well over 30 percent). When the dust has all settled (and even 150 years on, not all of Walker's has settled), it's likely that Alex will have named more *valid* species than Walker, and quite possibly more than anybody else.

Where was Mabel in all this? Simply put, she was everywhere. Mabel organized and indexed the crane fly collection so that Alex could quickly find material for comparison. (That was no small task, as the Alexanders' collection came to include tens of thousands of specimens representing around 13,000 different species, plus 50,000 or more microscope slides holding dissected genitalia and other structures.) She transcribed, typed, copyedited, proofread, and submitted manuscripts, and handled correspondence with journals about publishing them. She maintained and indexed their vast files of written material, including not just books and papers about flies but also Alex's voluminous correspondence and a huge collection of notes and papers about the history of entomology and the biographies of entomologists. On their summer collection trips, she drove every single mile (Alex never learned to drive), and she also joined Alex in collecting insects by day and sorting and processing them in the evenings. There was, it seems, little about the process of studying crane flies in which Mabel wasn't intimately involved. And, whenever Alex attended a scientific seminar, met with graduate students at his home, or invited a colleague to visit and talk about entomology, Mabel was there too: usually standing quietly or making small talk, but always there.

Here's where we can choose to tell, or to hear, one of two different stories about Mabel. Mabel working away for sixty years in support of her husband's scientific career suggests one story, as does Mabel standing quietly at Alex's side. It's a story rooted in the days when women had few opportunities for rewarding, independent careers in science, or for that matter in most other sectors of society. There's a long and sorry history of women providing unpaid and uncredited labor to support the careers of their husbands, and it's easy to place Mabel's work in that context. Especially easy, perhaps, given that Mabel's contribution began with, and always included, secretarial assistance: typing, filing, organizing, transcribing. These are skills that were, and perhaps still are, widely seen as "women's work" and devalued as menial; although it's not clear if women's careers were steered that way because the work was devalued, or if the causality ran the other way around. Mabel's devotion to the crane fly work was near total. The couple never had children, and Mabel doesn't seem to have taken on any other work after leaving her job at the Illinois Survey to marry Alex and follow him to his first job in Kansas. For most of Alex's career, too, he made little formal, or public, acknowledgment of Mabel's role. It was a time when such support from women was common and unremarkable, and Alex may well have simply slid into an arrangement that common custom made easy, and that helped him enormously in pursuing his passion for the crane flies. So that's Mabel's first story: subservient helpmate, unfairly relegated to stand behind her husband, unpaid and uncredited.

Accepting that first story without reflection, though, would greatly simplify something that may not have been so simple. It would be unfair to Mabel, and also to Alex. So let's consider a second version of Mabel's story. In this version, Mabel's partnership with Alex offered her a life she found more interesting and more rewarding than the vocation her secretarial training had prepared her for. Society expected her to take business dictation and file invoices, at least until she had children and left her job to care for them. Instead, she spent her life curating the world's best collection of crane flies and traveling the continent to collect and discover new species. Perhaps Mabel discovered that she loved and excelled at organizing and sorting and cataloguing (and perhaps she had even chosen secretarial training for that reason). Taxonomy, among all the sciences, offers those things in spades. That's not to diminish taxonomy as a science—quite the opposite. Taxonomy is nothing less than the attempt to discover the organization of life on Earth, an organization that reflects billions of years of evolutionary history. It's easy to imagine Mabel discovering that by joining in Alex's work she could tackle a problem bigger, more exciting, and more important than anything offered by secretarial school in 1910. In this version of Mabel's story, she was a collaborator, not a helpmate, taking an active and ever-expanding role in scientific aspects of the work: collecting, sorting, and preparing insects, curating the collection, drafting papers. In this version, she stood quietly behind Alex not because she was signalling subservience but simply because she was (according to those who knew her) always a very reserved person. In this version, she was essential to the science, a practicing entomologist who we should credit not just with typing, but with discovery.

This second version of Mabel's story credits her with the agency that the first story denies: not (or at least, not only) a woman relegated by societal convention to the helpmate's role, but one able to choose to play an active role in science. It's a story that sees Mabel subverting societal expectations for the path from secretary to wife and mother, working around the constraints she faced to find fulfillment—even if her path wasn't what we'd recognize now as a conventional, professional one. It sees her devoting her life to something she loved, along-side the person she loved, and never mind what society thought.

So which story is true? Both, I think, and therefore neither. It's probably true that Mabel, as the fourth of five children in a small-town family in the early 1900s, had limited opportunities. Secretarial training would have offered the prospects of income; a graduate degree in science, had she wanted it, would have been a very difficult reach. That means it's very likely that Mabel's path into science, via her marriage to Alex, was one scripted by the limited opportunities then available to women. And in the beginning, her contributions to the Alexanders' crane fly work may have been predominantly secretarial. Over the six decades of their marriage and work, though, it's clear that she became more and more a partner, more and more critical to the success of the enterprise.

There's evidence for the importance of Mabel's role from several sources. Since we're here by route of eponymous names, we can begin with the fourteen species (at least) that Alex named for Mabel: *Atarba margarita, Ctenophora margarita, Discobola margarita, Erioptera margarita, Hexatoma margaritae, Molophilus margarita, Pedicia margarita, Perlodes margarita, Phacelodocera margarita, Protanyderus margarita, Ptilogyna margaritae, Rhabdomastix margarita, Symplecta mabelana,* and *Neophrotoma mabelana.* Two of these use Mabel's first name, while the other twelve use "Margarita," Alex's affectionate nickname for her (derived from her middle name, Marguerite). Of course, scientists name species for their spouses all the time, and it's often an expression of love rather than a recognition of scientific contribution. But in this case, Alex recorded in his papers that he had both reasons for the naming. Early in his career, the names' dedications often recognized Mabel's help with collecting. For example, the stonefly *Perlodes margarita,* named in 1936, was collected "high up on the southeast side of Mount Washington, New Hampshire," and Alex wrote that he took "great pleasure in naming this interesting stone-fly in honor of my wife, Mabel Marguerite Alexander, who collected the type specimen and a host of other new and rare insects in many parts of the United States and Canada."[2] Later, though, the dedications recognized Mabel's broader contributions: for example, in his 1978 naming of *Molophilus margarita,* in which he wrote, "This distinct species is named in honor of my dear wife and life long collaborator in the study of the World Tipulidae."[3] Alex's namings *did* recognize Mabel as a scientist—although she could be his "dear wife" at the same time.

The suggestion from namings that Alex began to realize, and recognize, Mabel's contributions later in his career is backed up by another pattern. Alex's 1,000-plus papers were nearly all single-authored, but beginning in 1967 there's a series of eight papers on which Mabel receives co-authorship credit. They're chapters of two massive compendia: *A Catalogue of the Diptera* [flies] *of the Americas South of the United States,* and *A Catalogue of the Diptera of the Oriental Regions.* The Alexanders' chapters, together, span 500 pages and detail over 6,000 species of flies; they represent something of a synthesis of their joint crane fly work. The co-authorship acknowledges that it *was* joint work. Chris Thompson, who was a graduate student of Alex's in the 1960s (and who much later would name *Cepa margarita*), suggests that these Catalogues represent a dawning awareness on Alex's part: "Alex had been driven like everybody else in the academic world, you know, publish or perish; but all of a sudden he realized, especially in the Neotropical Catalogue, that the bottom line was Mabel did all the work!"[4]

Early on, perhaps, Alex took Mabel's contributions somewhat for granted. He acknowledged them in naming flies, but the main currency of credit among scientists—authorship—he didn't share. Whether he came to regret this we'll never know, but in the paper naming *Cepa margarita,* Thompson noted that late in Alex's career he "proclaimed to all that he could not have achieved his record of publication and new taxa described without his faithful teammate, Mabel."[5]

Alex made one last acknowledgment of Mabel's role, his most forceful and most poignant, after she passed away in September 1979. Chris Thompson went to Amherst for her funeral, and found Alex sitting in a chair staring into space. "It's over, Chris," he said. "Come get the collection."[6] Without Mabel, there would be—there *could* be—no more crane fly work.

As so often happens in life, we'll probably never know whether one of Mabel's two stories is closer to the truth. I like to think, though, that the juxtaposition of the two stories tells a larger one: that Alex's gradual realization that Mabel deserved credit as a full scientific partner reflects the (slow) progress of science in broadening and deepening the participation of women. If Alex and Mabel were high-school students today, there's no doubt some of the story would play out rather differently. Mabel would have many more educational and career options; Alex would likely assume less about the role of a spouse. And just maybe, they'd meet at the Survey not as scientist and secretary but as professional peers. But perhaps the science would play out the same—except with their 1,000 papers co-authored from the beginning.

What about Mabel's hoverfly, *Cepa margarita?* It wasn't named by Alex, but rather his student Chris Thompson—18 years after Alex's death, and 20 years after Mabel's. Thompson named two species at once, actually: *Cepa margarita* and *Cepa alex* (he originally used the genus name *Xela,* but that name turned out to already belong to a genus of fossil trilobites, and so under the rules of nomenclature couldn't be assigned to a fly.) Thompson wanted to honor his graduate adviser, Alex, but also to honor Mabel, and to recognize the equal partnership he saw in their work. So he chose a genus with two species, and named one for each. They're very similar species, and I think that matters: neither is bigger than the other, or brighter, or more widespread, or more important. *Cepa alex* has dark antennae, while *Cepa margarita*'s are orange, and there are some subtle differences in shading and venation of the wings; but otherwise, they're a matched pair. As Thompson explains, "the two included species are dedicated to Alex and Mabel Alexander, the most productive team of systematists ever. They described close to 11,000 new species, including some 10,000 crane flies. . . . Their publications total more [than] 1,017 titles, totaling over 20,000 pages. . . . Alex . . . personally proclaimed that he could not have achieved his record of publication and new taxa described without his faithful teammate, Mabel Margarita. So, we dedicate this genus to the Alexanders and the two included species to each team member."[7]

Cepa margarita's name, then, is Thompson's proclamation of Mabel's importance. It doesn't make up for the fact that Mabel's name *wasn't* on the first 1,000 papers, or for the fact that, in 1910, her career path was constrained by societal expectations. It's not a huge step toward equal opportunity, or equal recognition, for women in science. But it's a small step; and our path forward in science surely should involve both large steps and small ones to make sure that everyone can contribute, and that everyone who contributes is recognized. *Cepa margarita* does the latter in its small way, in remembering and honoring Mabel, and in calling attention to her story—or rather, to her two stories.

Cepa margarita and *Cepa alex:* Mabel's fly and Alex's. Mabel and Alex were together in a life of science that spanned six decades of social change. In *Cepa,* they're together forever, married in a genus of flower flies.

Madame Berthe's Mouse Lemur

We end our voyage, now, where we began it: in the tropical deciduous forests of western Madagascar. Here, as the rains bring relief in December after nine long months of drought, a soft chittering call from the branches may betray the presence of our smallest primate relative: Madame Berthe's mouse lemur, or *Microcebus berthae*. Why *berthae*? The lemurs, of course, don't care what name they bear. Their concerns as they rustle through the understory are more immediate: securing food, avoiding predators, finding mates. But you and I can wonder.

The name *Microcebus berthae,* it turns out, was coined in 2000 by Rodin Rasoloarison, Steve Goodman, and Jörg Ganzhorn. These three primatologists—one Malagasy, one American, and one German—were working together to puzzle out the diversity of *Microcebus* mouse lemurs living in the forest of western Madagascar. We now know that at least two dozen distinct mouse lemur species share Madagascar's forests, differing from each other in coloration, morphology, behavior, and geographic distribution. Rasoloarison and his colleagues, in their 2000 paper, distinguished seven species of *Microcebus* from western Madagascar. Three of these were newly recognized and received their scientific names: *Microcebus tavaratra, Microcebus sambiranensis,* and *Microcebus berthae.* Two of the species names refer to geography: *tavaratra* means "from the north" in the Malagasy

Madame Berthe's mouse lemur,
Microcebus berthae

language, while *sambiranensis* refers to that species' range in the Sambirano region of northwestern Madagascar. The third, *Microcebus berthae,* is eponymous, honoring a woman named Berthe Rakotosamimanana (or Madame Berthe, as she was generally called).

Dr. Berthe Rakotosamimanana (1938–2005) isn't a name that's famous around the world. Indeed, outside of Madagascar or the world of primatology research, few people are likely to have heard of her at all. But her name fits comfortably on the lemur that carries it, because she played a pivotal role in our appreciation of the rich biodiversity of Madagascar's fauna and in our efforts to conserve it.

Berthe Rakotosamimanana was born in Andasibé, a mining and farming village in the eastern rainforests of Madagascar. Madagascar was then a French colony. The colonial education system had begun, at the turn of the twentieth century, with clear dual tracks: an elite school system for French citizens and a very few well-positioned Malagasy children, and an Indigenous system designed to train everyone else as productive workers without offering much other opportunity. However, after the Second World War, the educational system was reformed to increase opportunity for Malagasy youth, and Berthe took advantage. After public school, she completed an undergraduate science degree and then went to France to pursue her doctorate in biological anthropology at the Université de Paris. In 1967, she returned to her home country (now independent from France) to teach at the Université d'Antananarivo. During a long career there, she founded first a paleontology unit, and 20 years later a Department of Paleontology and Biological Anthropology. In 31 years at the university, she taught thousands of Malagasy students working toward undergraduate science degrees. That alone would be a legacy worthy of respect, but Madame Berthe was at the center of much more.

Madagascar, like much of Africa, had a troubled post-colonial history. Recoiling from its French past, in the 1970s Madagascar descended into a long period of political tumult and found itself with a Marxist government under strong Soviet-bloc influence. The country became very isolated and very poor. Poor, that is, economically: it was increasingly obvious that Madagascar was home to an astonishingly rich diversity of life, much of it unstudied and most of it in urgent need of protection. Unfortunately, both science and conservation were very difficult there. The country's poverty meant that Malagasy scientists lacked the resources to tackle either the scientific questions or the conservation problems, while its political isolation made it extremely difficult for foreign scientists or conservation organizations to come to Madagascar to work. It's against this backdrop that we need to see, and measure, Madame Berthe's contribution.

So much scientific progress, and so much conservation work, depended on Madame Berthe that it's difficult to overstate her importance. In the Marxist era, the government of Madagascar was highly suspicious of outsiders, and it was nearly impossible for European or North American researchers to get permits to enter the country or to do research. Even taking photographs was fraught with bureaucracy: a permit from the Ministry of Culture was needed before a researcher could leave the country with pictures taken in Madagascar. Madame

Berthe made it possible for foreign researchers—notably primatologists, but many more besides—to come to Madagascar and study its fauna. She was the essential contact who could arrange research permits and use her connections in government to ease travel into, and around, the country. She could introduce researchers to local collaborators, land managers, or field assistants, and she could make it possible for research specimens to cross the country's borders. She never asked for coauthorship, or anything else, from these foreign researchers, so she might not turn up as having formal credit for the work, but she was essential nonetheless.

Without Madame Berthe, there's no doubt that *foreign* science in Madagascar would have been close to impossible. But her most important role, and the source of her real legacy, was in training young Malagasy scientists and conservationists. Not all the undergraduates who passed through her department or sat in her classes at the University of Madagascar went on to careers in science or conservation, of course, but a remarkable number did. Beyond that, she trained dozens of students in graduate degrees, mostly in primatology. She made a point of connecting local students with foreign researchers, and vice versa, so that both could benefit from international perspectives. Her students, both undergraduate and graduate, are now conservation leaders all over Madagascar; and many of them have made scientific contributions of their own.

Madame Berthe didn't teach just the technical science of biological anthropology, but also the ethos of science and of conservation. One of her Ph.D. students, Jonah Ratsimbazafy, puts it this way: "For us, she was not just a teacher, but Madame Berthe is also like our mother. She taught us to become responsible and guided us . . . to become a good citizen. Her dream was to build a professional training center in Madagascar . . . in order to ensure the survival of this unique treasure of Madagascar, the lemurs. . . . The seeds that she planted never stop growing."[1] In 1994, Madame Berthe founded the Groupe d'étude et de recherche sur les primates, or GERP (the Primate Research and Study Group). GERP is now the leading force in lemur research and conservation in Madagascar, and Jonah Ratsimbazafy is its secretary general.

Madame Berthe, in short, was the guiding light for a generation of Malagasy researchers and conservationists, and she was the fulcrum on which, for decades, foreign contributions to research and training in Madagascar turned. In naming *Microcebus berthae*, Rasoloarison, Goodman, and Ganzhorn were recognizing all this. Goodman and Ganzhorn are among the foreign scientists who couldn't have worked in Madagascar without Madame Berthe, and Rasoloarison is another of the Malagasy students she trained. Here's how they explained their choice of *berthae* for their new lemur species: "Madame Berthe, as she is known by hundreds of foreign researchers who have worked on Madagascar over the past 25 years and literally thousands of Malagasy students who have sought their degrees at the Université d'Antananarivo, has been one of the major forces in the advancement of Malagasy zoology, and in particular primatology."[2] Madame Berthe seems perfect for the mouse lemur, and the mouse lemur perfect for her.

Rasoloarison and his colleagues took great pleasure in naming a lemur for Madame Berthe, but by all accounts Madame Berthe took even greater pleasure in having one named for her. She talked about it often, with a mixture of pride, delight, and self-deprecation. She was a short woman, but stout, and seemed particularly amused that *M. berthae* is the smallest of all living primates. Colleagues remember her pointing to photographs of her lemur, saying things like, "You know, the smallest primate in the world," then trailing off in a pregnant pause while looking down toward her own feet. She was a woman with a prestigious university appointment, close government connections, and a lot of power, but given an opportunity to poke fun at herself, she didn't hesitate. The situation for *Microcebus berthae* in the wild is, unfortunately, not so amusing. Its populations, and their chittering in the dusk, are endangered and declining. About 8,000 adults are thought to remain, in a handful of forests spread over less than 500 square kilometers (just one sixth the size of Rhode Island, that tiniest of U.S. states). Thirteen other *Microcebus* species are similarly (or more seriously) endangered, and four more are vulnerable. Their plight is a common one in Madagascar. Across the country, which remains one of the world's poorest, people scrape subsistence lives from the landscape. Forests are being burned, cleared, and fragmented; soils are being degraded and eroded so rivers run red with silt to the sea; animals are overharvested for food and for the global pet trade; and invasive species such as Asian toads, tilapia, and guava trees drive native counterparts from their habitats.

Extinction, in Madagascar, isn't just a hypothetical: a wave of extinctions followed human colonization about 2,000 years ago. Among the earliest victims were spectacular megafauna, including giant sloth lemurs (the largest, at 160 kilograms, weighing nearly as much as a gorilla) and elephant birds (the largest weighing 700 kilograms and standing 3 meters tall). Both were gone, likely hunted to extinction, about 1,000 years ago. Dozens of other species, large and small, followed. These have included, of course, some eponymously named creatures. Consider, for example, a bird in the cuckoo family: the snail-eating coua, *Coua delalandei*. It was named in 1827 by the Dutch zoologist Coenraad Temminck, with the name honoring Pierre-Antoine Delalande. Delalande (1787–1823) was a French naturalist who made a two-year collecting expedition to southern Africa. He sent over 19,000 African specimens back to the Paris Museum of

Natural History. Among these specimens were 2,000 birds, 10,000 insects, 6,000 plants, and a complete skeleton of a 75-foot southern right whale, which Delalande spent two months dissecting from a washed-up carcass despite what he called its "appalling and infectious stench."[3] Oddly, while Temminck credits Delalande with collecting the coua that bears his name, this seems unlikely. Delalande collected a lot of birds, to be sure, but the coua inhabited only the northeastern Madgascar island of Nosy Baraha, and his report on his South African expeditions makes no mention of a visit there. Perhaps he would eventually have visited Madagascar, and seen his coua, but his life was cut short: just three years after returning from South Africa, he died, in his beloved museum, at the age of 37. The coua's life was cut short too. By 1850, deforestation, overhunting, and the introduction of rats to Nosy Baraha had driven the species to extinction, and Delalande is now honored only in bones and skin. Madame Berthe, as of this book's writing, is still honored in living flesh, in eyeshine and chittering calls in the forest; but there's no guarantee that her mouse lemur won't follow Delalande's snail-eating coua into extinction.

No guarantee, perhaps, but there is hope. Madagascar's biodiversity has a high public profile. Ecotourism has come to play an important role in the country's economy, providing both financial resources and incentives for preserving natural areas and the creatures that inhabit them. Conservation dollars are flowing into the country, as conservation organizations worldwide work on behalf of Madame Berthe's mouse lemur and its compatriots. In Madagascar, too, grassroots conservation organizations—many of them led, or staffed, by Malagasy conservationists trained by Madame Berthe or by her students—are active across the country. Scientific work continues, with Malagasy and international scientists in collaboration. Again, Madame Berthe's influence persists: many of the Malagasy partners in this research were trained by her, and many of the international partners began their Madagascar work under her sponsorship. As a result of all this research, new species are discovered in Madagascar every year, and more is known about the species we've already recognized. Both parts of that are critical, because attempts to protect species without knowledge of their habitats and distribution and behavior are doomed to flail in the dark. Madagascar's landscape and its flora and fauna are still critically imperiled, but we know what needs to be done, and we've begun to do it. If we succeed—if *Microcebus berthae* still chitters from the forest at dusk in years and in centuries to come—it will be in no small part because of Madame Berthe's passion and energy and the legacy of research, and of researchers, she left to us.

There's a lot behind *Microcebus berthae,* then. Its name tells a story: of Madame Berthe and her legacy, and her connections to "her" mouse lemur and to the scientists—Malagasy and Western—who named it. It's a story that can be told whenever the name is used; and that's a wonderful thing, because Madame Berthe's story deserves the telling. It has good company. Thousands of stories lie behind thousands of eponymous Latin names; this book has only scratched the surface. From Sid Vicious to Richard Spruce; from William Spurling to Charles Darwin; from Conrad Isberg to Prasanna Dharmapriya; from Pierre-Antoine Delalande to Berthe Rakotosamimanana: eponymous names weave the world together. In doing so, they reveal much not just about the organisms that bear the names and the people they celebrate, but also about the thinking, and the personalities, of the scientists who coin them.

Ignominy and heroism; obscurity and fame; hostility and love; loss and hope. It's all in a Latin name.

Notes

Preface

1. Quoted in Kastner 1977:24.

Chapter 1. The Need for Names

1. Grothendieck 1986:24, my translation.

Chapter 3. *Forsythia, Magnolia,* and Names Within Names

1. Plumier 1703, my translation.
2. Magnol 1689, translation from Stearn 1961.

Chapter 4. Gary Larson's Louse

1. D. Clayton, conversation with S. Heard via Skype, May 18, 2017.
2. Clayton 1990:260.
3. Larson 1989:171.
4. Fisher et al. 2017:399.
5. N. Shubin, phone conversation with S. Heard, June 22, 2018.
6. Quoted in Kelley 2012:24.

Chapter 5. Maria Sibylla Merian and the Metamorphosis of Natural History

1. Quoted in Todd 2007:173.
2. Nahakara et al. 2018:285.

Chapter 6. David Bowie's Spider, Beyoncé's Fly, and Frank Zappa's Jellyfish

1. Lessard and Yeates 2011:247.
2. Lessard and Yeates 2011:248.

3. Quoted by Murkin n.d., and confirmed by Boero, personal communication.
4. Quoted by Murkin n.d., and confirmed by Boero, personal communication.

Chapter 8. The Name of Evil

1. Scheibel 1937:440.
2. Linnaeus 1737, *Critica Botanica,* translation from Hort 1938:71.
3. Fischer et al. 2014:62.
4. Geilis 2011:638.
5. Miller and Wheeler 2005:126.
6. Cuvier 1798:71.
7. Letter from Louis Agassiz to Elizabeth Agassiz, quoted in Gould 1980:173.

Chapter 9. Richard Spruce and the Love of Liverworts

1. Spruce 1908, I:95.
2. Spruce 1908, I:364.
3. Spruce 1861:10.
4. Spruce 1861:12.
5. Letter of 1873, quoted in Spruce 1908:xxxix.
6. Spruce 1908, I:140.
7. Letter of 1873, quoted in Spruce 1908:xxxix.

Chapter 10. Names from the Ego

1. Hort 1938:64.
2. Anonymous 1865:181.
3. Sabine 1818:522.
4. Angas 1849 Plate XXIX.
5. Jamrach 1875:2.
6. Jamrach 1875:2.
7. Tytler 1864:188, emphasis in original.

Chapter 11. Robert von Beringe's Gorilla and Dian Fossey's Tarsier

1. Von Beringe, F. R. 1903. "Bericht des Hauptmanns von Beringe über seine Expedition nach Ruanda" (Report of Captain von Beringe on his expedition to Rwanda). *Deutsches Kolonialblatt* (quoted in Schaller 1963:390).
2. Niemitz et al. 1991:105.

3. D. Fossey, letter to I. Redmond, July 1976; quoted in de la Bédoyère 2005.

Chapter 12. Less Than a Tribute

1. Linnaeus 1729, *Praeludia Sponsaliorum Plantarum,* translation from Larson 1967.

2. Linnaeus 1737, *Critica Botanica;* translation from Hort 1938:63.

3. Isberg 1934:263, my translation.

4. Peterson 1905:212.

5. K. Miller, email to S. Heard, July 7, 2017.

6. Nazari 2017:89.

7. V. Nazari, email exchange with S. Heard, July 10, 2017.

Chapter 13. Charles Darwin's Tangled Bank

1. Wulf 2015:8.

2. G. Monteith, email to S. Heard, August 6, 2018.

3. Darwin 1859:490.

Chapter 14. Love in a Latin Name

1. Bonaparte 1854:1075, my translation.

2. Lesson 1839:44, my translation.

3. Gross 2016.

4. Miller and Wheeler 2005:89.

5. Hamet 1912, my translation.

6. Haeckel 1899–1904, translation from Richards 2009a.

7. Quoted in, and translation from, Richards 2009b.

8. Quoted in, and translation from, Blunt 1971.

Chapter 15. The Indigenous Blind Spot

1. Markle et al. 1991:284.

2. Le Vaillant 1796:208.

3. Layard 1854:127.

4. Quoted in Papa 2012:93.

Chapter 16. Harry Potter and the Name of the Species

1. Mendoza and Ng 2017:26.

2. Ahmed et al. 2016:25.

Chapter 17. Marjorie Courtenay-Latimer and the Fish from the Depths of Time

1. Quoted in Weinberg 2000:11.
2. Quoted in Weinberg 2000:2.
3. Smith 1956:27.
4. Smith 1956:41.
5. Quoted in Weinberg 2000:27.
6. Quoted in Weinberg 2000:19.
7. Woodward 1940:53.
8. Smith 1939b:749–50.

Chapter 18. Names for Sale

1. Köhler et al. 2011:219.
2. Quoted in Chang 2008.
3. Evangelista 2014.
4. The Boeing Company, n.d. General Information, http://www.boeing.com /company/general-info, accessed April 7, 2019.

Chapter 19. A Fly for Mabel Alexander

1. Anonymous 1874:140.
2. Alexander 1936:24, 27.
3. Alexander 1978:268.
4. F. C. Thompson, telephone interview with S. Heard, October 17, 2018; transcript on file.
5. Ibid.
6. Ibid.
7. Thompson 1999:341.

Epilogue

1. Jonah Ratsimbazafy, email to S. Heard, November 2, 2018.
2. Rasoloarison et al. 2001:1004.
3. Delalande 1822:5.

Sources and Further Reading

Preface

Byatt, A. S. 1994. "Morpho Eugenia," in *Angels and Insects*. New York: Vintage Books.

Kastner, Joseph. 1977. *A Species of Eternity*. New York: Alfred A. Knopf.

Introduction

Rasoloarison, Rodin M., Steven M. Goodman, and Jörg U. Ganzhorn. 2000. "Taxonomic revision of mouse lemurs (*Microcebus*) in the western portions of Madagascar." *International Journal of Primatology* 21, no. 6: 963–1019.

Yoder, Anne D., David W. Weisrock, Rodin M. Rasoloarison, and Peter M. Kappeler. 2016. "Cheirogaleid diversity and evolution: Big questions about small primates." In *The Dwarf and Mouse Lemurs of Madagascar: Biology, Behavior and Conservation Biogeography of the Cheirogaleidae*, edited by Shawn M. Lehman, Ute Radespiel, and Elke Zimmermann, 3–20. Cambridge: Cambridge University Press.

Chapter 1

Giller, Geoffrey. 2014. "Are we any closer to knowing how many species there are on Earth?" *Scientific American*. Springer Nature. April 8, 2014. https://www.scientificamerican.com/article/are-we-any-closer-to-knowing-how-many-species-there-are-on-earth/

Grothendieck, Alexander. 1985. *Récoltes et semailles: Réflexions et témoignages sur un passé de mathématicien* [Harvests and sowings: Reflections and testimonies on a mathematician's past]. Université des Sciences et Techniques du Languedoc, Montpellier, et Centre National de la Recherche Scientifique. Accessed May 31, 2017. lipn.univ-paris13.fr/~duchamp /Books&more/Grothendieck/RS/pdf/RetS.pdf.

Johnson, Kristin. 2012. *Ordering Life: Karl Jordan and the Naturalist Tradition.* Baltimore: Johns Hopkins University Press (especially with respect to trinomials, Chapter 2, "Reforming Entomology" and Chapter 3, "Ordering Beetles, Butterflies, and Moths").

Lewis, Daniel. 2012. *The Feathery Tribe: Robert Ridgway and the Modern Study of Birds.* New Haven: Yale University Press (especially with respect to trinomials, Chapter 6, "Publications about Birds").

Moss, Stephen. 2018. *Mrs. Moreau's Warbler: How Birds Got Their Names.* London: Faber and Faber.

Stearn, William Thomas. 1959. "The background of Linnaeus's contributions to the nomenclature and methods of systematic biology." *Systematic Zoology* 8, no. 1: 4–22.

Wright, John. 2014. *The Naming of the Shrew: A Curious History of Latin Names.* London: Bloomsbury Publishing.

Chapter 2

Burkhardt, Lotte. 2016. *Verzeichnis Eponymischer Pflanzennamen* [Index of Eponymyic Plant Names]. Botanic Garden and Botanical Museum Berlin.

Figueiredo, Estrela, and Gideon F. Smith. 2010. "What's in a name: Epithets in *Aloe* L. (Asphodelaceae) and what to call the next new species." *Bradleya* 28: 79–102.

Lewis, Daniel. 2012. *The Feathery Tribe: Robert Ridgway and the Modern Study of Birds.* New Haven: Yale University Press (Chapter 5, "Nomenclatural Struggles, Checklists, and Codes").

Turland, Nicholas J., John H. Wiersema, Fred R. Barrie, Werner Greuter, David L. Hawksworth, Patrick S. Herendeen, Sandra Knapp, Wolf-Henning Kusber, De-Zhu Li, Karol Marhold, Tom W. May, John Mc-

Neill, Anna M. Monro, Jefferson Prado, Michelle J. Price, and Gideon F. Smith (eds.). 2018. *International Code of Nomenclature for Algae, Fungi, and Plants (Shenzhen Code) Adopted by the Nineteenth International Botanical Congress Shenzhen, China, July 2017.* Regnum Vegetabile 159. Glashütten: Koeltz Botanical Books. https://doi.org/10.12705/Code.2018

International Commission on Zoological Nomenclature. 1999. "International Code of Zoological Nomenclature, 4th ed." *The International Trust for Zoological Nomenclature.* http://www.nhm.ac.uk/hosted-sites/iczn/code/

Winston, Judith E. 1999. *Describing Species: Practical Taxonomic Procedure for Biologists.* New York: Columbia University Press.

Chapter 3

Aiello, Tony. 2003. "Pierre Magnol: His life and works." *Magnolia, the Journal of the Magnolia Society* 38, no. 1: 1–10.

Dulieu, Louis. 1959. "Les Magnol" [The Magnols]. *Revue d'histoire des sciences et de leurs applications* 12, no. 3: 209–224.

Magnol, Petrus. 1689. *Prodromus Historiae Generalis Plantarum in quo Familiae Plantarum per Tabulas Disponuntur* [Precursor to a General History of Plants, in Which the Families of Plants Are Arranged in Tables]. Montpellier.

Plumier, P. Carolo. 1703. *Nova Plantarum Americanarum Genera* [New Genera of American Plants]. Paris.

Smith, Archibald William. 1997. *A Gardener's Handbook of Plant Names: Their Meanings and Origins.* Mineola, New York: Dover Publications. (*Forsythia:* pp. 160–161)

Stearn, William Thomas. 1961. "Botanical gardens and botanical literature in the eighteenth century." In *Catalogue of Botanical Books in the Collection of Rachel McMasters Miller Hunt.* 2(1), edited by Allan Stevenson, xli–cxl. Pittsburgh: The Hunt Botanical Library.

Treasure, Geoffrey. 2013. *The Huguenots.* New Haven: Yale University Press.

Chapter 4

Barrowclough, George F., Joel Cracraft, John Klicka, and Robert M. Zinck. 2016. "How many kinds of birds are there and why does it matter?" *PLoS ONE* 11:e0166307.

Clayton, Dale H. 1990. "Host specificity of *Strigiphilus* owl lice (Ischnocera: Philopteridae), with the description of new species and host associations." *Journal of Medical Entomology* 27, no. 3: 257–265.

Fisher, J. Ray, Danielle M. Fisher, Michael J. Skvarla, Whitney A. Nelson, and Ashley P. G. Dowling. 2017. "Revision of torrent mites (Parasitengona, Torrenticolidae, *Torrenticola*) of the United States and Canada: 90 descriptions, molecular phylogenetics, and a key to species." *ZooKeys* 701: 1.

Kelley, Theresa M. 2012. *Clandestine Marriage: Botany and Romantic Culture.* Baltimore: Johns Hopkins University Press.

Kohler, Robert E. 2006. *All Creatures: Naturalists, Collectors, and Biodiversity, 1850–1950.* Princeton: Princeton University Press (Chapter 6, "Knowledge").

Larson, Gary. 1989. *The Prehistory of the Far Side.* Kansas City: Andrews and McMeel Publishing.

Chapter 5

Davis, Natalie Z. 1995. *Women on the Margins: Three Seventeenth-Century Lives.* Cambridge: Harvard University Press.

Merian, Maria Sibylla. 1675. *Neues Blumenbuch* [New Book of Flowers]. Nuremburg: Johann Andreas Graffen.

———. 1679. *Der Raupen Wunderbare Verwandlung und Sonderbare Blumennahrung* [The Wonderful Transformation of Caterpillars and Their Remarkable Diet of Flowers]. Nuremberg and Frankfurt: Andreas Knorz, for Johann Andreas Graff and David Funck.

———. 1705. *Metamorphosis Insectorum Surinamensium* [Metamorphosis of the Insects of Suriname]. Amsterdam: Gerard Valk.

Nakahara, Shinichi, John R. Macdonald, Francisco Delgado, and Pablo Sebastián Padrón. 2018. "Discovery of a rare and striking new pierid butterfly from Panama (Lepidoptera: Pieridae)." *Zootaxa* 4527, no. 2: 281–291.

Rücker, Elisabeth. 2000. *The Life and Personality of Maria Sibylla Merian.* Preface to *Metamorphosis Insectorum Surinamensium.* Pion, London: Pion Press reprint.

Stearn, William Thomas. 1978. "Introduction," in *The Wondrous Transformation of Caterpillars: 50 Engravings Selected from Erucarum Ortus,* by Maria Sybilla Merian (1718). London: Scolar Press.

Todd, Kim. 2007. *Chrysalis: Maria Sibylla Merian and the Secrets of Metamorphosis.* New York: I.B. Taurus.

Chapter 6

Boero, Fernando. 1987. "Life cycles of *Phialella zappai* n. sp., *Phialella fragilis* and *Phialella sp.* (Cnidaria, Leptomedusae, Phialellidae) from central California." *Journal of Natural History* 21, no. 2: 465–480.

Jäger, Peter. 2008. "Revision of the huntsman spider genus *Heteropoda* Latreille 1804: Species with exceptional male palpal conformations (Araneae: Sparassidae: Heteropodinae)." *Senckenbergiana Biologica* 88, no. 2: 239–310.

Lessard, Bryan D., and David K. Yeates. 2011. "New species of the Australian horse fly subgenus *Scaptia* (*Plinthina*) Walker 1850 (Diptera: Tabanidae), including species descriptions and a revised key." *Australian Journal of Entomology* 50, no. 3: 241–252.

Murkin, Andy. (n.d.) "Here's your jelly, Frank!" Andymurkin dotcom. Accessed February 28, 2017. www.andymurkin.net/Resources/MusicRes/ZapRes /jellyfish.html (confirmed via personal correspondence with F. Boero).

Chapter 7

Anonymous. 1854. "Supposed murder of a portion of the passengers and crew of the ketch Vision." *Moreton Bay Free Press,* November 21, 1854; reprinted, *Sydney Morning Herald,* November 29, p. 4.

Angus, George French. 1874. *The Wreck of the "Admella," and Other Poems.* London: Sampson, Low, Marston & Searle.

Australian National Herbarium. "Strange, Frederick (1826–1854)." Council of Heads of Australasian Herbaria, Biographical Notes. Accessed February 4, 2017. www.anbg.gov.au/biography/strange-frederick.html.

Comben, Patrick. 2018. *The Mysteries of Frederick Strange, Naturalist.* Brisbane: Self-published.

New South Wales. Parliament. Legislative Council. 1855. *Search by H.M.S. Ship "Torch" for the survivors of the "Ningpo," and for the remains of the late Mr Strange and his companions.* Sydney: New South Wales Legislative Council.

Gee, Jane, et al. 2008–2015. *Murdered in Australia 10.1854.* British Genealogy. Accessed February 4, 2017. Thread of posts: www.british-genealogy. com/threads/25880-murdered-in-Australia-10.1854.

Iredale, Tom. 1933. "Systematic notes on Australian land shells." *Records of the Australian Museum* 19, no. 1: 37–59.

———. 1937. "A basic list of the land Mollusca of Australia.—Part II." *Australian Zoologist* 9, no. 1: 1–39.

Kloot, Tess. 1983. "Iredale, Tom (1880–1972)." In *Australian Dictionary of Biography. Volume 9, 1891–1939, Gil–Las,* edited by Bede Nairn, Geoffrey Serle and Chris Cunneen. Melbourne: Melbourne University Press.

Kohler, Robert E. 2006. *All Creatures: Naturalists, Collectors, and Biodiversity, 1850–1950.* Princeton: Princeton University Press (Chapter 4, "Expedition").

MacGillivray, John, George Busk, Edward Forbes, and Adam White. 1852. *Narrative of the Voyage of HMS Rattlesnake, Commanded by the Late Captain Owen Stanley, R.N., F.R.S. &c. During the Years 1846–1850. Including Discoveries and Surveys in New Guinea, the Louisiade Archipelago, etc. to which is Added the Account of Mr. E.B. Kennedy's Expedition for the Exploration of the Cape York Peninsula.* London: T. & W. Boone.

Meston, Archibald. 1895. *Geographic History of Queensland.* Brisbane: E. Gregory, Government printer.

Morgan, E.J.R. 1966. "Angas, George French (1822–1886)." In *Australian Dictionary of Biography. Volume 1, 1788–1850, A–H,* edited by Douglas Pike. Melbourne: Melbourne University Press. Accessed online February 4, 2017. http://adb.anu.edu.au/biography/angas-george-french-1708 /text1857.

Noonan, Patrick. 2016. "Sons of Science: Remembering John Gould's Martyred Collectors." *Australasian Journal of Victorian Studies* 21, no. 1: 28–42.

van Wyhe, John. 2018. "Wallace's Help: The many people who aided AR Wallace in the Malay Archipelago." *Journal of the Malaysian Branch of the Royal Asiatic Society* 91(1), no. 314: 41–68.

Whitley, Gilbert Percy. 1972. "The life and work of Tom Iredale (1880–1972)." *Australian Zoologist* 17, no. 2: 65–125.

Whittell, Hubert Massey. 1941. "Frederick Strange: A biography." *Australian Zoologist* 11: 96–114.

Chapter 8

Cuvier, George. 1798. *Tableau Elementaire de l'Histoire Naturelle des Animaux* (Elementary Table of the Natural History of Animals). Paris: Baudouin, imprimeur.

Fischer, Valentin, Maxim S. Arkhangelsky, Gleb N. Uspensky, Ilya M. Stenshin, and Pascal Godefroit. 2014. "A new Lower Cretaceous ichthyosaur from Russia reveals skull shape conservatism within Ophthalmosaurinae." *Geological Magazine* 151, no. 1: 60–70.

Gielis, Cees. 2011. "Review of the Neotropical species of the family Pterophoridae, part II: Pterophorinae (Oidaematophorini, Pterophorini) (Lepidoptera)." *Zoologische Mededelingen* 85: 589.

Gould, Stephen J. 1980. *The Panda's Thumb: More Reflections in Natural History* (Chapter 16, "Flaws in a Victorian veil"). New York: Norton.

Guthörl, Paul. 1934. "Die Arthropoden aus dem Carbon und Perm des-Saar-Nahe-Pfalz-Gebietes" [Carboniferous and Permian age arthropods from the Saar-Nahe region]. *Abhandlungen der Preußischen Geologischen Landesanstalt* (N.F.) 164: 1–219.

Hort, Arthur. 1938. *The "Critica Botanica" of Linnaeus*. London: Ray Society.

Menand, Louis. 2001. "Morton, Agassiz, and the origins of scientific racism in the United States." *The Journal of Blacks in Higher Education* 34: 110–113.

Miller, Kelly B., and Quentin D. Wheeler. 2005. "Slime-mold beetles of the genus *Agathidium* Panzer in North and Central America, Part II. Coleoptera: Leiodidae." *Bulletin of the American Museum of Natural History* 291: 1–167.

Reidel, Alexander, and Raden Pramesa Narakusumo. 2019. "One hundred and three new species of *Trigonopterus* weevils from Sulawesi." *ZooKeys* 828: 1–153.

Scheibel, Oskar. 1937. "Ein neuer *Anophthalmus* aus Jugoslawien" [A New Anophthalmus from Yugoslavia]. *Entomologische Blätter* 33, no. 6: 438–440.

Chapter 9

Ayers, Elaine. 2015. "Richard Spruce and the trials of Victorian bryology." *The Public Domain Review.* October 14, 2015. publicdomainreview. org/2015/10/14/richard-spruce-and-the-trials-of-victorian-bryology/

Gribben, Mary, and John Gribben. 2008. *The Flower Hunters.* Oxford: Oxford University Press. (Chapter 8, "Richard Spruce")

Honigsbaum, Mark. 2003. *The Fever Trail: In Search of the Cure for Malaria.* London: Pan Macmillan.

Seward, M.R.D., and S.M.D. FitzGerald (eds.). 1996. *Richard Spruce (1817–1893): Botanist and Explorer.* London: Royal Botanic Gardens, Kew.

Spruce, Richard. 1861. *Report on the Expedition to Procure Seeds and Plants of the Cinchona succirubra, or Red Bark Tree.* London: Her Majesty's Stationery Office.

———. 1908. *Notes of a Botanist on the Amazon & Andes,* edited by A. R. Wallace. London: MacMillan.

Chapter 10

Angas, George French. 1849. *The Kafirs Illustrated in a Series of Drawings Taken Among the Amazulu, Amaponda, and Amakosa tribes.* London: J. Hogarth.

Anonymous. 1865. "Malacologie d'Algérie (review)." *American Journal of Malacology* 1, no. 181.

Blunt, Wilfrid. 1971. *The Compleat Naturalist: A Life of Linnaeus,* introduction by William T. Stearn. London: Collins.

Bourguignat, Jules René. 1864. *Malacologie de l'Algerie, ou Histoire Naturelle des Animaux Mollusques Terrestres et Fluviatiles Recueillis jusqu'à ce Jour dans nos Possessions du Nord de l'Afrique* [Malacology of Algeria, or Natural History of Terrestrial and Fluvial Molluscs Collected to this Day in Our Possessions from North Africa]. Paris: Challamel Ainé.

Caleb, John T. D. 2017. "Jumping spiders of the genus *Icius* Simon, 1876 (Araneae: Salticidae) from India, with a description of a new species." *Arthropoda Selecta* 26, no. 4: 323–327.

Cartwright, Oscar Ling. 1967. "Two New Species of *Cartwrightia* from Central and South America (Coleoptera: Scarabaeidae: Aphodiinae)." *Proceedings of the United States National Museum* 124, no. 3632: 1–8.

Dance, S. Peter. 1968. "J. R. Bourgignat's *Malacologie de l'Algérie.*" *Journal of the Society for the Bibliography of Natural History* 5, no. 1: 19–22.

Farber, Paul L. 2000. *Finding Order in Nature: The Naturalist Tradition from Linnaeus to E. O. Wilson.* Baltimore: Johns Hopkins University Press. (Chapter 1, "Collecting, Classifying, and Interpreting Nature," on Linnaeus's vanity)

Hort, Arthur. 1938. *The "Critica Botanica" of Linnaeus.* London: Ray Society.

Jamrach, William. 1875. "On a new species of Indian rhinoceros." Reproduced in *The Rhinoceros in Captivity* by L. C. Rookmaaker. The Hague: SPB Academic Publishing, 1998.

Linnaeus, Carl. 1729. *Spolia Botanica.* Handwritten manuscript. The Linnean Society of London, *The Linnaean Collections online.* http://linnean-online.org/61284/

———. 1730. *Fundamenta Botanica.* Handwritten manuscript. The Linnean Society of London, *The Linnaean Collections Online.* http://linnean-online.org/61328/

Sabine, Joseph. 1818. "An account of a new species of gull lately discovered on the west coast of Greenland." *Transactions of the Linnean Society of London* 12: 520–523.

Sanderson, Ivan Terence. 1937. *Animal Treasure.* New York: Viking Press.

Spangler, Paul J. 1985. "Oscar Ling Cartwright, 1900–1983." *Proceedings of the Entomological Society of Washington* 87, no. 3: 690–692.

Tytler, Robert Christopher. 1864. "Description of a new species of *Paradoxurus* from the Andaman Islands." *Journal of the Asiatic Society of Bengal* 33, no. 294: 188.

Wright, John. 2014. *The Naming of the Shrew: A Curious History of Latin Names.* Bloomsbury Publishing.

Chapter 11

de la Bédoyère, Camilla. 2005. *No One Loved Gorillas More: Dian Fossey, Letters from the Mist.* Vancouver: Raincoast Books.

Greenbaum, Eli. 2017. *Emerald Labyrinth: A Scientist's Adventures in the Jungles of the Congo.* Lebanon, N.H.: University Press of New England. (Chapter 2, on Rudolf Grauer)

Hayes, Harold T. P. 1990. *The Dark Romance of Dian Fossey.* New York: Simon and Schuster.

Mowat, Farley. 1987. *Virunga: The Passion of Dian Fossey.* Toronto: McClelland and Stewart.

Niemitz, C., A. Nietsch, S. Warter, and Y. Rumpler. 1991. "*Tarsius dianae:* A new primate species from Central Sulawesi (Indonesia)." *Folia Primatologica* 56, no. 2: 105–116.

Schaller, George B. 1963. *The Mountain Gorilla: Ecology and Behaviour.* Chicago: University of Chicago Press.

Shekelle, Myron, Colin P. Groves, Ibnu Maryanto, and Russell A. Mittermeier. 2017. "Two new tarsier species (Tarsiidae, Primates) and the biogeography of Sulawesi, Indonesia." *Primate Conservation* 31: 61–69.

Stapleton, Timothy J. 2017. *A History of Genocide in Africa.* Santa Barbara, Calif.: Praeger.

Chapter 12

Greuter, Werner. 1976. "The flora of Psara (E. Aegean Islands, Greece)—an annotated catalogue." *Candollea* 31: 191–242.

Gribben, Mary, and John Gribben. 2008. *The Flower Hunters.* Oxford: Oxford University Press. (Chapter 1, "Carl Linnaeus")

Hort, Arthur. 1938. *The "Critica Botanica" of Linnaeus.* London: Ray Society.

Isberg, O. 1934. *Studien über Lamellibranchiaten des Leptaenakalkes in Dalarna: Beitrag zu einer Orientierung über die Muschelfauna im Ordovicium und Silur* [Studies on Lamellibranchiates of the Leptaena Lime-

stone in Dalarne: Contribution to a Guide to the Mussel Fauna in Or-
dovician and Silurian periods]. Lund: Häkan Ohlssons Buchdruckerei.

Larson, James L. 1967. "Linnaeus and the natural method." *Isis* 58, no. 3: 304–320.

Linnaeus, Carl. 1729. *Praeludia Sponsaliorum Plantarum.* Thesis.

Miller, Kelly B., and Quentin D. Wheeler. 2005. "Slime-mold beetles of the genus *Agathidium* Panzer in North and Central America, Part II. Coleoptera:
Leiodidae." *Bulletin of the American Museum of Natural History* 291: 1–167.

Nazari, Vazrick. 2017. "Review of *Neopalpa* Povolný, 1998 with description
of a new species from California and Baja California, Mexico (Lepidoptera, Gelechiidae)." *ZooKeys* 646: 79.

Peterson, O. A. 1905. "Preliminary note on a gigantic mammal from the
Loup Fork beds of Nebraska." *Science* 22, no. 555: 211–212.

Siegebeck, Johann G. 1736. *Letter to C. Linnaeus, 28 December 1736* (in Latin).
Letter. The Linnaean Correspondence. http://linnaeus.c18.net/Letter
/L0119.

Sloan, Robert E. 1996. *The Autobiography of Robert Evan Sloan.* Unpublished. https://studylib.net/doc/13045745/autobiography-of-robert-
evan-sloan. Accessed July 8, 2017.

Stevens, Peter F. 1994. *The Development of Biological Systematics: Antoine-
Laurent de Jussieu, Nature, and the Natural System.* New York: Columbia
University Press.

Warburg, Elsa. 1925. "The trilobites of the Leptæna limestone in Dalarne:
With a discussion of the zoological position and the classification of the
Trilobita." *Bulletin of the Geological Institute of Uppsala* XVII.

Chapter 13

Beccaloni, George. 2008. "Plants and animals named after Wallace." *The
Alfred Russel Wallace Website.* January 12, 2008. http://wallacefund.info
/plants-and-animals-named-after-wallace.

Bushnell, Mark. 2017. "A Vermonter's life in plants remembered." *Vermont
Daily Digger,* July 2, 2017. https://vtdigger.org/2017/07/02/a-vermonters-
life-in-plants-remembered.

Darwin, Charles. 1859. *On the Origin of Species by Means of Natural Selection, or the Preservation of Favoured Races in the Struggle for Life.* London: John Murray,

Miličić, Dragana, Luka Lučić, and Sofija Pavković-Lučić. 2011. "How many Darwins?—List of animal taxa named after Charles Darwin." *Natura Montenegrina* 10, no. 4: 515–532.

Oberprieler, Rolf, Christopher Lyal, Kimberi Pullen, Mario Elgueta, Richard Leschen, and Samuel Brown. 2018. "A Tribute to Guillermo (Willy) Kuschel (1918–2017)." *Diversity* 10, no. 3: 101.

Wulf, Andrea. 2015. *The Invention of Nature: Alexander von Humboldt's New World.* New York: Vintage Books.

Chapter 14

Araújo, João P. M., Harry C. Evans, Ryan Kepler, and David P. Hughes. 2018. "Zombie-ant fungi across continents: 15 new species and new combinations within *Ophiocordyceps.* I. Myrmecophilous hirsutelloid species." *Studies in Mycology* 90: 119–160.

Blunt, Wilfrid. 1971. *The Compleat Naturalist: A Life of Linnaeus,* introduction by William T. Stearn. London: Collins.

Bonaparte, Charles-Lucien Prince. 1854. "Coup d'oeil sur les Pigeons (deuxième parti)" [A glance at the pigeons (part two)]. *Comptes Rendus Hebdomadaires des Séances de l'Académie des Sciences* 39: 1072–1078.

Figueiredo, Estrela, and Gideon F. Smith. 2010. "What's in a name: Epithets in *Aloe* L. (Asphodelaceae) and what to call the next new species." *Bradleya* 28: 79–102.

Finsch, Otto. 1902. "Ueber zwei neue Vogelarten von Java" [On two new birds from Java]. *Notes from the Leyden Museum* 23, no. 3: 147–152.

Gross, Rachel E. 2016. "How newly discovered species get their weird names." *Slate.* January 25, 2016. www.slate.com/articles/health_and_science/science/2016/01/how_newly_discovered_species_get_names_from_taxonomists.html.

Haeckel, Ernst. 1899–1904. *Kunstformen der Natur* [Art Forms in Nature]. Leipzig: Verlag des Bibliographischen Instituts.

Hamet, M. Raymond. 1912. "Sur un nouveau *Kalanchoe* de la baie de Delagoa" [On a New *Kalanchoe* from Delagoa Bay]. *Repertorium Novarum Specierum Regni Vegetabilis* 11, no. 16–20: 292–294.

Huang, Chih-Wei, Yen-Chen Lee, Si-Min Lin, and Wen-Lung Wu. 2014. "Taxonomic revision of *Aegista subchinensis* (Möllendorff, 1884) (Stylommatophora, Bradybaenidae) and a description of a new species of *Aegista* from eastern Taiwan based on multilocus phylogeny and comparative morphology." *ZooKeys* 445: 31–55.

Lesson, René Primevère. 1839. "Oiseaux rares ou nouveaux de la collection du Docteur Abeillé, à Bordeaux" [Rare or New Birds from Dr. Abeillé's Collection in Bordeaux]. *Revue Zoologique par La Société Cuvierienne* 2: 40–43.

Miller, Kelly B., and Quentin D. Wheeler. 2005. "Slime-mold beetles of the genus *Agathidium* Panzer in North and Central America, Part II. Coleoptera: Leiodidae." *Bulletin of the American Museum of Natural History* 291: 1–167.

Pensoft Editorial Team. 2014. "A new land snail species named for equal marriage rights." *Pensoft blog.* Pensoft. October 13, 2014. https://blog.pensoft.net/2014/10/13/a-new-land-snail-species-named-for-equal-marriage-rights.

Richards, Robert J. 2009a. "The tragic sense of Ernst Haeckel: His scientific and artistic struggles." In *Darwin: Art and the Search for Origins,* edited by Pamela Kort and Max Hollein, 92–103. Frankfurt: Schirn-Kunsthalle Gallery.

———. 2009b. *The Tragic Sense of Life: Ernst Haeckel and the Struggle over Evolutionary Thought.* Chicago: University of Chicago Press. (Especially Chapter 10, "Love in a Time of War")

Velmala, Saara, Leena Myllys, Trevor Goward, Håkon Holien, and Pekka Halonen. 2014. "Taxonomy of *Bryoria* section *Implexae* (Parmeliaceae, Lecanoromycetes) in North America and Europe, based on chemical, morphological and molecular data." *Annales Botanici Fennici* 51, no. 6: 345–371.

Chapter 15

Ascherson, Paul F. A. 1880. "Ueber die Veränderungen, welche die Blüthen-hüllen bei den Arten der Gattung *Homalium* Jacq. nach der Befruchtung erleiden und die für die Verbreitung der Früchte von Bedeutung zu sein scheinen" [On Variation in Flowers of the Genus *Homalium* Jacq. After Fertilization, and Its Importance for Fruit Distribution]. *Sitzungsberichte der Gesellschaft Naturforschender Freunde zu Berlin* 1880, no. 8: 126–133.

Berlin, Brent. 1992. *Ethnobiological Classification: Principles of Categorization of Plants and Animals in Traditional Societies.* Princeton: Princeton University Press.

Clarke, Philip A. 2008. *Aboriginal Plant Collectors: Botanists and Australian Aboriginal People in the Nineteenth Century.* Kenthurst, Australia: Rosenberg.

Doty, Maxwell S. 1978. "*Izziella abbottae,* a new genus and species among the gelatinous Rhodophyta." *Phycologia* 17, no. 1: 33–39.

Figueiredo, Estrela, and Gideon F. Smith. 2010. "What's in a name: epithets in *Aloe* L. (Asphodelaceae) and what to call the next new species." *Bradleya* 28: 79–102.

Glaskin, K., M. Tonkinson, Y. Musharbash, and V. Burbank (eds.). 2008. *Mortality, Mourning, and Mortuary Practices in Indigenous Australia.* Burlington, Vt.: Ashgate.

Layard, Edgar Leopold. 1854. "V.—Notes on the ornithology of Ceylon, collected during an eight years' residence in the island." *Annals and Magazine of Natural History* 14, no. 79: 57–64.

Le Vaillant, François. 1796. *Travels into the Interior Parts of Africa, by Way of the Cape of Good Hope; in the Years 1780, 81, 82, 83, 84, and 85.* Vol. I, 2nd edition. London: G.G. and J. Robinson.

Markle, Douglas F., Todd N. Pearsons, and Debra T. Bills. 1991. "Natural history of *Oregonichthys* (Pisces: Cyprinidae), with a description of a new species from the Umpqua River of Oregon." *Copeia* 1991, no. 2: 277–293.

Nicholas, George. 2018. "It's taken thousands of years, but Western science is finally catching up to traditional knowledge." *The Conversation,* February 14, 2018. https://theconversation.com/its-taken-thousands-of-years-

but-western-science-is-finally-catching-up-to-traditional-knowl
edge-90291.

Papa, J. W. 2012. "The Appropriate Use of Te Reo Māori in the Scientific
Names of New Species Discovered in Aotearoa New Zealand." M.Sc.
thesis, University of Waikato, Hamilton.

Seldon, David S., and Richard A. B. Leschen. 2011. "Revision of the *Meco-
dema curvidens* species group (Coleoptera: Carabidae: Broscini)." *Zootaxa* 2829: 1–45.

Thornton, Thomas F. 1997. "Anthropological studies of Native American
place naming." *American Indian Quarterly* 21, no. 2: 209–228.

van Wyhe, John. 2018. "Wallace's Help: The many people who aided AR
Wallace in the Malay Archipelago." *Journal of the Malaysian Branch of
the Royal Asiatic Society* 91, no. 1: 41–68.

Whaanga, Hēmi, Judy Wiki Papa, Priscilla Wehi, and Tom Roa. 2013. "The
use of the Māori language in species nomenclature." *Journal of Marine
and Island Cultures* 2, no. 2: 78–84.

Wood, Hannah M., and Nikolaj Scharff. 2017. "A review of the Madagascan
pelican spiders of the genera *Eriauchenius* O. Pickard-Cambridge, 1881
and *Madagascarchaea* gen. n. (Araneae, Archaeidae)." *ZooKeys* 727: 1–96.

Chapter 16

Ahmed, Javed, Rajashree Khalap, and J. N. Sumukha. 2016. "A new species
of dry foliage mimicking *Eriovixia* Archer, 1951 from Central Western
Ghats, India (Araneae: Araneidae)." *Indian Journal of Arachnology* 5,
no. 1–2: 24–27.

Barbosa, Diego N., and Celso O. Azevedo. 2014. "Revision of the Neotropical *Laelius* (Hymenoptera: Bethylidae) with notes on some Nearctic species." *Zoologia (Curitiba)* 31, no. 3: 285–311.

Butcher, B. Areekul, M. Alex Smith, Mike J. Sharkey, and Donald LJ
Quicke. 2012. "A turbo-taxonomic study of Thai *Aleiodes* (*Aleiodes*) and
Aleiodes (*Arcaleiodes*) (Hymenoptera: Braconidae: Rogadinae) based
largely on COI barcoded specimens, with rapid descriptions of 179 new
species." *Zootaxa* 3457, no. 1: 232.

Hauer, Tomáš, Marketa Bohunicka, and Radka Muehlsteinova. 2013. "*Calochaete gen. nov.* (Cyanobacteria, Nostocales), a new cyanobacterial type from the 'páramo' zone in Costa Rica." *Phytotaxa* 109, no. 1: 36–44.

Heller, John L. 1945. "Classical mythology in the *Systema Naturae* of Linnaeus." *Transactions and Proceedings of the American Philological Association* 76: 333–357.

Mendoza, Jose C. E., and Peter K. L. Ng. 2017. "*Harryplax severus,* a new genus and species of an unusual coral rubble-inhabiting crab from Guam (Crustacea, Brachyura, Christmaplacidae)." *ZooKeys* 647: 23–35.

Moratelli, Ricardo, and Don E. Wilson. 2014. "A new species of *Myotis* (Chiroptera, Vespertilionidae) from Bolivia." *Journal of Mammalogy* 95, no. 4: E17–E25.

Reidel, Alexander, and Raden Pramesa Narakusumo. 2019. "One hundred and three new species of *Trigonopterus* weevils from Sulawesi." *ZooKeys* 828: 1–153.

Saunders, Thomas E., and Darren F. Ward. 2017. "A new species of *Lusius* (Hymenoptera: Ichneumonidae) from New Zealand." *New Zealand Entomologist* 40, no. 2: 72–78.

Chapter 17

Anonymous. 2004. "Marjorie Courtenay-Latimer." Obituary. *The Telegraph* (London), May 19, 2004. https://www.telegraph.co.uk/news/obituaries/1462225/Marjorie-Courtenay-Latimer.html.

Courtenay-Latimer, M. 1979. "My story of the first coelacanth." *Occasional Papers of the California Academy of Science* 134: 6–10.

Smith, John L. B. 1939a. "A living fish of Mesozoic type." *Nature* 143, no. 3620: 455–456.

———. 1939b. "The living coelacanthid fish from South Africa." *Nature* 143, no. 3627: 748–750.

———. 1956. *Old Fourlegs: The Story of the First Coelacanth.* London: Longman, Green.

Thomson, Keith Stewart. 1991. *Living Fossil: The Story of the Coelacanth.* New York: Norton.

Weinberg, Samantha. 2000. *A Fish Caught in Time: The Search for the Coelacanth.* New York: HarperCollins.

Woodward, Arthur Smith. 1940. "The surviving crossopterygian fish, *Latimeria.*" *Nature* 146, no. 3689: 53–54.

Chapter 18

Carbayo, Fernando, and Antonio C. Marques. 2011. "The costs of describing the entire animal kingdom." *Trends in Ecology & Evolution* 26, no. 4: 154–155.

Chang, Alicia. 2008. "Immortality all in a name." *The Toronto Star,* July 5, 2008. https://www.thestar.com/life/2008/07/05/immortality_all_in_a_name.html.

Evangelista, Dominic. 2014. "Vengeful taxonomy: Your chance to name a new species of cockroach." *Entomology Today.* The Entomological Society of America. March 20, 2014. entomologytoday.org/2014/03/20/vengeful-taxonomy-your-chance-to-name-a-new-species-of-cockroach.

Johnson, Kristin. 2012. *Ordering Life: Karl Jordan and the Naturalist Tradition.* Baltimore: Johns Hopkins University Press (see "Conclusion," on biodiversity inventories).

Köhler, Jörn, Frank Glaw, Gonçalo M. Rosa, Philip-Sebastian Gehring, Maciej Pabijan, Franco Andreone, Miguel Vences, and H. L. Darmstadt. 2011. "Two new bright-eyed treefrogs of the genus *Boophis* from Madagascar." *Salamandra* 47, no. 4: 207–221.

Montanari, Shaena. 2019. "Taxonomy for Sale to the Highest Bidder." *Undark,* April 10, 2019. https://undark.org/article/nomenclature-auctions-bidder.

Trivedi, Bijal P. 2005. "What's in a species' name? More than $450,000." *Science* 307: 1399.

Wallace, Robert B., Humberto Gómez, Annika Felton, and Adam M. Felton. 2006. "On a new species of titi monkey, genus *Callicebus* Thomas (Primates, Pitheciidae), from western Bolivia with preliminary notes on distribution and abundance." *Primate Conservation* 231, no. 36: 29–40.

Wilson, Edward O. 1985. "The biological diversity crisis: A challenge to science." *Issues in Science and Technology* 2: 20–29.

Chapter 19

Alexander, Charles P. 1936. "A new species of *Perlodes* from the White Mountains, New Hampshire (Family Perlidae; Order Plecoptera)." *Bulletin of the Brooklyn Entomological Society* 31: 24–27.

———. 1978. "New or little-known Neotropical Tipulidae (Diptera). II." *Transactions of the American Entomological Society* 104, no. 3: 243–273.

Alexander, Charles P., and Mabel M. Alexander. 1967. "Family Tanyderidae." Chapter 5 in *A Catalogue of the Diptera of the Americas South of the United States,* 1–3. São Paulo: Departamento de Zoologia, Secretaria da Agricultura.

Anonymous. 1874. "Francis Walker." Obituary. *The Entomologist's Monthly Magazine* 11: 140–141.

Byers, G. W. 1982. "In memoriam: Charles P. Alexander, 1889–1991." *Journal of the Kansas Entomological Society* 55: 409–417.

Dahl, C. 1992. "Memories of crane-fly heaven." *Acta Zoologica Cracoviensia* 35, no. 1: 7–9.

Gurney, A. B. 1959. "Charles Paul Alexander." *Fernald Club Yearbook* (Fernald Entomology Club, University of Massachusetts) 28: 1–6.

Knizeski, H. M., Jr. 1979. "Dr. Charles Paul Alexander." *Journal of the New York Entomological Society* 87: 186–188.

Ohl, Michael. 2018. *The Art of Naming.* Translated by Elisabeth Lauffer. Cambridge: MIT Press. (C. P. and Mabel Alexander, pp. 191–195)

Thompson, F. Christian. 1999. "A key to the genera of the flower flies (Diptera: Syrphidae) of the Neotropical Region including descriptions of new genera and species and a glossary of taxonomic terms used." *Contributions on Entomology, International* 3, no. 3: 321–348.

Epilogue

Barnard, Keppel H. 1956. "Pierre-Antoine Delalande, naturalist, and his Cape visit, 1818–1820." *Quarterly Bulletin of the South African Library* 11, no. 1: 6–10.

Delalande, M. P. 1822. *Précis d'un Voyage au Cap de Bonne Ésperance, Fait par Ordre du Gouvernement* [Summary of a Voyage to the Cape of Good Hope, Made by Government Order]. Paris: A. Belin.

Rasoloarison, Rodin M., Steven M. Goodman, and Jörg U. Ganzhorn. 2000. "Taxonomic revision of mouse lemurs (*Microcebus*) in the western portions of Madagascar." *International Journal of Primatology* 21, no. 6: 963–1019.

Acknowledgments

A legion of friends and colleagues helped with this book. Some read and commented on drafts. Some helped with research, supplied sources, or answered questions about obscure points of history or taxonomy. Some kindly let me interview them about Latin names they'd coined or about species named for them. Some helped me translate sources in German, Italian, Latin, Swedish, Russian, and other languages I read only with great difficulty. Still others suggested names I absolutely had to cover (and to those whose suggestions didn't find a place in the book, I apologize).

No list I can offer will be complete; but among those who deserve my thanks are Richard Willan, Priantha Wigesinghe, Hēmi Whaanga, Alexa Alexander Trusiak, Adrian Tronson, Eric Tong, Nic Tippery, Tom Thornton, Michaela Thompson, Chris Thompson, Adam Summers, John Stanisic, Alex Smith, Neil Shubin, David Shorthouse, Yana Shibel, Catherine Sheard, Manu Saunders, Gary Saunders, Charles Sacobie, Rebekah Rogers, Leigh Anne Riedman, David Rider, Julian Resasco, Jonah Ratsimbazafy, Dinesh Rao, Sasanka Ranasinghe, Amy Parachnowitsch, Robert Owens, Michael Orr, Patrick Noonan, Kärin Nickelsen, Vazrick Nazari, Staffan Müller-Wille, Peter Moonlight, Arne Mooers, Geoff Monteith, Julia Mlynarek, Russ Mittermeier, Kelly Mitchell, Kelly Miller, Rainer Melzer, Jack McLachlan, Bill Mattson, Karl Magnacca, Wayne Maddison, Dan Lewis, Hans-Walter Lack, Jörn Köhler, Frank Köhler, Jitka Klimešová, Lewis Kelly, Niklas Janz, John Huisman, Christie Henry, Steve Hendrix, Rebecca Helm, Kristie Heard, Jamie Heard, Malorie Hayes, Mikael Gyllström, Steve Goodman, Donna Giberson, Mischa Giasson, Ann Marie Gawel, Jörg Ganzhorn,

227

Jannice Friedman, David Frank, Graham Forbes, Lesley Fleming, Zen Faulkes, Neal Evenhuis, Dominic Evangelista, Emily Damstra, Les Cwynar, Doug Currie, Julie Cruikshank, Pat Comben, Dale Clayton, Toni Carmichael, Alexssandro Camargo, George Byers, Doug Byers, Mike Bruton, Fenja Brodo, Alex Bond, Fernando Boero, Jason Bittel, Claus Bätke, Victor Baranov, Tony Aiello, and Javed Ahmed.

I'm also grateful to many Twitter followers and blog readers—far too many to list here—for comments and answers to some of the questions I've broadcast there. Staff at the UNB Libraries, especially the Document Delivery section, were an enormous help tracing even the most obscure publications for my research. (I once asked a staff member at my university library whether I'd win a contest for ordering the campus's weirdest variety of interlibrary loans. "Um," she said, and left a pregnant pause before adding, "We're not supposed to notice.")

At the Yale University Press, Jean Thomson Black, Michael Deneen, and Phil King patiently answered my questions and steered the manuscript toward being a book. A Grant in Aid of Scholarly Book Publishing from the Harrison McCain Foundation provided resources in support of this project. Finally, Kristie and Jamie Heard were remarkably patient with me through the long process of writing this book (and my previous one, for that matter). Thank you.

Index

INDEX

Torch, HMS, 62
Torrenticola shubini, 37
traditional knowledge, 145
Tragelaphus angasii, 90
Trewavas, Ethelwynn, 91–92
Trigonopterus spp., 70, 156
Trigonopterus watsoni, 70
trilobites: *Arcticalymene* spp., 68–69;
  Han solo, 156; *Isbergia* spp., 108
trinomials, 11–13
Trochilus franciae, 130, 131
Trump, Donald, 111–113
twinflower (*Linnaeus borealis*), 85–89
Tytler, Robert, 93–94

Umpqua chub (*Oregonicthys kalawatseti*), 141–142
Université d'Antananarivo, 195
Uslar-Gleichen, Frida von, 135–137

Vader, Darth, 111
Vahl, Martin, 25–26
Vanessa tameamea, 142
van Wyhe, John, 144
variation, 12–13, 15–17, 33–34
Vedda people (Sri Lanka), 141
Vengeful Taxonomy auction, 178–179
Vicious, Sid, 68
Victoria, Queen, 143
Victoria amazonica, 143
Vlasimsky, Stan, and family, 175

Walker, Francis, 185
walking shark (*Hemiscyllium galei*), 177
Wallace, Alfred Russel, 82, 102,
  118–119, 126–127, 144
Warburg, Elsa, 108–109
Warburgia spp., 108–109

wasps, 152–153; *Aleiodes* spp., 155;
  Aleiodes stewarti, 51, 56; *Diolcogaster ichiroi,* 51–52; *Laelius* spp., 155; *Lusius malfoyi,* 153; *Polemistus chewbacca,*
  156; Tolkien's dwarves, 157
water lily (*Victoria amazonica*), 143
waterweed (*Ledermanniella maturiniana*), 155
Watson, James, 70
Wayag Island, 177
Weberbauer, Augusto, 122
weevils: *Eupholus vlasimskii,* 175;
  Macrostyphlus spp., 157; *Trigonopterus* spp., 156; *Trigonopterus watsoni,* 70
whale (*Livyatan melvillei*), 52
Wheeler, Quentin, 110–111, 133–134
white mulberry (*Morus alba*), 20
willow (*Homalium abdessammadii*),
  146
Wilson, E. O., 180
Winslet, Kate, 52
Winston, Judith, 132
Wonderful Transformation of Caterpillars and Their Remarkable Diet of Flowers, The (Merian), 41
Wulf, Andrea, 121
Wyeomyia smithii, 56

Xestoblatta berenbaumae, 178–179
Xiphophorus nezahualcoyotl, 142

Yeates, David, 51
Your Inner Fish (Shubin), 36

Zappa, Frank, 49, 55
Zenaida spp., 132
Zoological Code, 18–22, 85, 113